Arnaud Stéphane R. Tapsoba

West African cattle: diversity and genetic structure

Arnaud Stéphane R. Tapsoba

West African cattle: diversity and genetic structure

Characterization and genetic structure of West African cattle

ScienciaScripts

Imprint

Cover image: www.ingimage.com

This book is a translation from the original published under ISBN 978-620-6-72271-7.

Publisher:
Sciencia Scripts
is a trademark of
Dodo Books Indian Ocean Ltd. and OmniScriptum S.R.L publishing group

120 High Road, East Finchley, London, N2 9ED, United Kingdom
Str. Armeneasca 28/1, office 1, Chisinau MD-2012, Republic of Moldova, Europe
Printed at: see last page
ISBN: 978-620-8-19828-2

Contents

DEDICATION

I dëdie this book:

To eternal God

Thank you Lord for all the regal graces. Thank you Lord for having me poЛë during moments of dëtresse. Thank you for all the moments of joy. Because the road is still a long way off, oh! Lord "hold my hand over the path". Yours is the reign, the power and the glory for siëcles and siëcles. Amen.

To my Father,

Thank you Dad for my ëducation. Thank you for your rigour. From a young age you taught me to work well and to persevere. With you I learnt that nothing can be taken for granted in life and that to live is to know how to face the dark moments of life with courage and determination.

To my Mother

To you the firstëre woman in my life Mum, thank you for your love and for your tenderness. Thank you for the gift of life. Thank you for all the times you came home tired and still helped me prepare my compositions. Thank you for putting up with my whims and my impatience all these years. Your advice has guided me throughout my studies. May this work be the culmination of all your sacrifices and efforts.

SUMMARY

West Africa is characterised by the existence of an important zoo gënëtic diversity characterised by the presence of trypanotolerant taurines (*Bos taurus*). However, the introduction of the Zebu (*Bos indicus*) to the continent has profoundly altered the genetic make-up of local breeds, particularly bulls. This is all the more worrying in that little characterisation work has been carried out on re-evaluating the levels of introgression of the region's bull breeds. The general objective is to carry out a zootechnical and molecular characterisation of the cattle breeds in Burkina Faso in comparison with those of certain West African countries. More specifically, the aim will be (i) to use morphological data to infer gene flows between types of cattle in Burkina Faso, (ii) to determine the levels of diversity and introgression of taurine populations in Burkina Faso and (iii) to assess the genetic structure of West African breeds, including those of Burkina Faso. Sampling for morphological characterisation was carried out in the provinces of Sanguie, Nahouri and Poni for bulls and in the provinces of Oudalan and Soum for zebus. The results of the morphological characterisation showed a strong structuring of the taurine populations in the South-West, Centre-West and Centre-South regions in 2014. The Linear Discriminant Analysis used for classification purposes and to estimate probable gene flows showed that in 2014, 97.69%, 100% and 98.7% of Gourounsi bulls from the Nahouri, Sanguie and Poni regions respectively, were well classified within their original population. In contrast, in 2018, a strong homogenisation signal was observed in the Gourounsi bullfighting populations. PCA revealed differentiation between Gourounsi and Lobi. The molecular study of the bull breeds in Burkina also revealed the close proximity between the Gourounsi bull sub-populations and their high level of introgression by Zebu (between 20 and 50%). However, genetic diversity exists within these populations, with the average heterozygosity observed being 0.64, 0.7 and 0.58 in Gourounsi taurines from Nahouri and Sanguie and Lobi from Poni respectively, while the expected overall heterozygosity was 0.67 (Gourounsi Nahouri); 0.71 (Gourounsi Sanguie) and 0.6 (Lobi). Similar to the morphological study, the molecular evaluation revealed little differentiation between the Gourounsi taurine subpopulations, with 2% of the differences attributed to differences between the subpopulations. Strong introgression of West African taurines by zebu gdnes was detected in the second molecular study, which looked at 12 cattle populations (Burkina Faso, Benin, Niger and Mali).

Keywords: *Biometric morphology, gene flow, introgression, taurines, zebu*

INTRODUCTION

The livestock of West Africa is characterised by great diversity, since it includes taurines (*Bos taurus*), zëbuses (*Bos indicus*), as well as numerous intermediate types resulting from more or less ancient and extensive interbreeding, linked to human migrations and pastoralism. The zëbus occupy the driest regions, such as the Sahel, while the taurins are found in the more humid areas, infested by tsetse flies, or "tsë-tsë", the main vectors of trypanosomes (Lhoste 1991). These taurines are in fact trvpanotolerant, which is a real advantage for these areas, given the difficulty of developing a vaccine and the disadvantages of chemical control. According to Epstein (1971), the first domestic cattle planted in Africa were humpless, long-horned animals. This group is represented by disparate types with different names (Bakosi, Baoule, Lagunaire, Muturu, Nuba, Somba) and also includes regional variants. Zebu represent a third, later contribution. These were animals with a thoracic hump, whose introduction into Africa seems to have begun around the fourth century AD, and whose spread intensified after the Arab invasion (669 AD). These various genetic types were widely mixed in certain areas. According to Epstein (1971), all African cattle come from the same region of origin, South-West Asia. Long-horned bulls were first domesticated there at the beginning of the fifth millennium BC, probably in Mesopotamia, from the local population of wild breeds. It was then from these long-horned taurines that the short-horned taurines were selected, probably in Elam, during the fourth millennium BC, as well as the zebus, from the same period, possibly in the semi-arid steppe bordering the east of the Great Dirty Desert of Iran. However, the work of Loftus *et al* (1994) on bovine mitochondrial DNA tends to call into question Epstein's theory on the origin of zebus. Indeed, analysis of the sequence differences observed between the mitochondrial DNA of taurines and that of Indian zebus suggests that the process of divergence between these two types of DNA dates back at least 200,000 years, and would therefore predate the Neolithic period of domestication. This would mean that *Bos taurus* and *Bos indicus* would have been domesticated from two distinct wild populations, with the original home of the zebus possibly being in present-day Pakistan. Curiously, all African zebus seem to possess both the acrocentric Y chromosome of the zebu and the mitochondrial DNA type of the taurine, suggesting that the introduction of zebu genes into Africa was essentially male (Lhoste, 1991). Furthermore, after identifying microsatellite alleles specific to Indian zëbu (alleles absent in European bulls), MacHugh *et al.* (1997) found that the relative frequency of these alleles in African zëbu populations increased from east to west, which appears to reflect the gradient of introduction of Indian zëbu into pre-existing bull populations. All these results confirm that the bovine settlement of Africa, first by long-horned taurines, then by short-horned taurines, and finally by zebus, was a highly complex process.

African cattle populations have been the subject of various descriptions and inventories, remarkably summarised in the monograph by Epstein (1971), as well as numerous studies of a zootechnical nature. However, the potential information provided by morpho-biometric measurements has yet to be fully exploited. The work carried out in the subregion using morphological parameters has made it possible to characterise a few West African cattle breeds, but has not highlighted the probable genetic relationships that may exist between the different cattle populations studied (Traore *et al.,* 2016; Grema *et al.,* 2017). The aim of this study is (1) to use morphological measurements to establish the different gene flows that may exist between the different bull breeds of Burkina Faso; (2) to establish the levels of genetic

diversity and introgression of the bull populations of Burkina Faso; and (3) to assess the genetic structure of West African breeds, including those of Burkina Faso.
This thesis is divided into four chapters. Chapter I reviews the literature and the various tools and methods for phenotypic and molecular characterisation, the ethnological and zootechnical characteristics of the local and imported cattle breeds of Burkina Faso and the different production systems, as well as the evolutionary history of the *Bovidae* family. Chapter II describes the equipment and methods used for the phenotypic and molecular characterisation of the local cattle breeds of Burkina Faso. Chapter III describes the results and Chapter IV gives the discussion, conclusion and perspectives.

CHAPTER I

BIBLIOGRAPHICAL REVIEW

1.1. Origin and domestication of cattle

Bosprimigenius is thought to have appeared in the Pleistocene in a region stretching from Turkestan to India and Arabia (Magin *et al.,* 2008). Its fossils are most numerous in the Siwalik region of India. From there, it spread throughout the Eurasian region and North Africa at the end of the Great Ice Age 250,000 years ago, while adopting numerous local forms that can be grouped into two types: the Indian *Bos primigenius namadicus*, the origin of the zebu, and the Western *Bos primigenius*, the origin of domestic cattle. The African form, *Bos primigenius opisthonomus/mauretanicus/africanus* is linked to the latter branch (Guintard *et al.,* 2008) of domestication (Edwards *et al.,* 2007). However, analysis of the sequences of the complete D-loop region and one of these hypervariable regions of several African, European and Asian breeds reveals two distinct lines, one grouping African cattle (taurines and zebus) and the other only Asian cattle (zebus) (Edwards *et al.,* 2007).

The origin and history of African cattle appear to be complex, as local breeds are often the result of a vast context of nomadic movements, pastoral migrations and successive introductions of Asian animals. In Africa, the first indigenous cattle breeds are exclusively considered to be taurines. These taurine populations are thought to have resulted from successive pastoral migrations from the Near East (MacHugh *et al.,* 1997). Recent tirclieological findings suggest that the African form of Aurochs (*Bosprimigenius opisthonomus)* was domesticated independently on the African continent (Wendorf and Schild, 1994; MacHugh *et al.,* 1997). This suggestion seems to be confirmed on the one hand by mtDNA analyses which reveal two distinct origins of domestication between African and European taurines (Bradley *et al.,* 1996); and on the other hand, by analysis of genetic distances from microsatellite data, which indicates that the gëne pools of African and Asian taurines diverged around 150 000 to 250 000 years ago (MacHugh *et al.,* 1997). In addition, contemporary African zebus are thought to have resulted from ancient interbreeding between the first zebus introduced to the African continent (via trade and migration between the Fertile Crescent and the Indian subcontinent) and indigenous taurine populations (Bradley *et al.,* 1994). From there, they spread through Arab migrations to northern and eastern Africa (Esptein, 1971). According to MacHugh *et al* (1997), the zebu Y chromosome haplotype may have reached high frequencies in breeds classified as pure taurines. In addition, analysis of microsatellite loci in breeds of African zebu including Gobra zebu and Maure zebu (Senegal), Asian zebu, African taurines including the N'Dama breed (Senegal, Gambia, Guinee Bissau and Guinea Bissau), and African taurine breeds (Senegal, Gambia, Guinee Bissau and Guinea Bissau) has shown that the Y chromosome haplotype of the zebu has a high frequency.

(thin) and e'uropean bullfighters, a r^ëlë 1 gënëtic introgression of African zëbus into West African bullfighting populations through the presence of diagnostic zëbu allies. The frequency of these specific allies of zëbu in bullfighting populations follows an increasing gradient from the north to the south of West Africa (MacHugh *et al.,* 1997).

1.2. Evolutionary history of the family *Bovidae* (*Artiodactyla, Mammalia*)

Cattle in the strict sense are animals of the genus *Bos,* which gave its name to the family *Bovidae* (Gray, 1821) or "carnivorous" ruminants. They include not only the taurine subspecies "*Bos taurus*", (Linnaeus, 1758), but also the zebu "*Bos taurus indicus*", (Linnaeus,

1758), the yak "*Bos grunniens*", (Linnaeus, 1758), the gayal or gaur "*Bos frontalis*" and the bateng "*Bos javanicus*", (Guintard *et al.*, 2008). In a broader sense, they can refer to animals belonging to the subfamily *Bovinae*. In a restrictive sense, the word bovine can sometimes refer only to cattle (*Bos taurus*) and exclude buffalo, yak, etc. (Guintard al., 2008). The family *Bovidae* (*Cetartiodactyla, Mammalia*) is one of the most diverse of the large mammals, with almost 140 species currently recognised (Guintard *et al.*, 2008).

1.2.1. Phyletic evolution of the Bovidae (*Artiodactyla, Mammalia*)

PhylogCnCtic analysis conducted on ribosomal gënes sequences (12S and 16S rRNA) shows that the *Bovidae* form a paraphyletic group with non-robust intertribal relationships, contrary to most analyses in which additional evolutionary steps are invoked to satisfy bovid monophyly (Gatesy *et al.*, 1992). In addition, several independent molecular studies have also hypothesised this paraphyly of bovids. Resolving the question of monophyly in the *Bovidae* has been a major task. Consequently, the estimation of the time of divergence of the different *Bovidae* lineages based on transversion-type substitutions (Tv) shows that the first cladogenesis, which occurred during the Lower Miocene at around 20 million years ago, produced two major groups of *Bovidae*: the *Bovinae* (*Bovini*, *Boselaphini* and *Tragelaphini*) and the *Antilopinae* (all the other extant tribes), which evolved in Eurasia and Africa respectively. This suggests that the *Bovinae* and *Antilopinae* underwent rapid, contemporary tribal radiation in the Middle Miocene, giving rise to four *Bovinae* lines (around 12-14.3 million years ago) and six *Antilopinae* lines (13.6-15.3 million years ago). This split of the *Bovidae* family into two main clades (bovine and non-bovine) was demonstrated by Arif *et al.* (2012) by studying the phylogenetic information provided by several 5
mitochondrial genes (12S rRNA, 16S rRNA, COI, Cyt-B and D-loop) compared to that of full-length mitochondrial DNA. Furthermore, the main phases of Involution in the *Bovidae* have been demonstrated by comparisons of Cytochrome B sequences, and as evidenced by paleontological data showing that the different groups of Bovidës underwent several migrations between Africa and Eurasia. As a result, the *Bovidae* family has turned out to be monoplivyletic, with two major clades: the *Bovinae* and the *Antilopinae*. Two contemporary periods of cladogenesis appear to have occurred during the Middle Miocene, giving rise to most of the extant tribes, which underwent subsequent radiation at the end of the Lower Miocene/Pliocene (Hassanin and Douzery, 1999).

1.2.2. Phyletic evolution of the *Bovinae* and its tribes (*Artiodactyla, Bovidae*)

Fossil data suggest a common South Asian origin for the subfamily *Bovinae*. The tribe of mammals *Bovini* (subfamily *Bovinae*, family *Bovidae*) contains the world's most important domesticated species. The various domestication processes of these animals are among the most significant advances of the Neolithic transition. With a specialised digestive system, these species can use cellulose as a source of energy, with advantages for the production of milk, meat and hides (Lenstra and Bradley, 1999). The split of the *Bovinae* has resulted in three main tribes (Table I). The first represents two clades of the *Bovini* and the other two those of the *Boselaphini* and *Tragelaphini*. The representatives of the latter two tribes are mainly hunted for their meat and skins.

Analysis of the phylogenetic relationships of the *Bovini* tribe and evidence of ancient polymorphisms inferred from the sequences of nucleus genes in relation to QTLs involved in milk production (located on chromosomes 1, 2, 4 and 9) showed an association between members of the *Bovinae* sub-tribes with strong divergence of the majority of lines. Estimation of divergence times using a relaxed molecular clock (calibrated from Bison and Yak fossils, 2 and 1.7 million years ago respectively), shows that the other members of the *Bovini* tribe diverged from *Bos taurus* around 2-3 million years ago. The Bubalina sub-tribe diverged from *Bos taurus* around 5-9 million years ago. These divergences fit well with the phylogenetic reconstructions derived from the Kimura 2-parameter model and the NeighborJoining algorithm in agreement with p-distances, revealing a separation of the Bubalina and Bovina subtribes as well as the genera *Bubalus* and *Syncerus*. The genetic grouping between Yak, *Bison* and *Bos taurus* may explain the similarity observed between the European *Bison* and *Bos taurus.*

(MacEachern *et al.,* 2009). Other studies have also shown that *Bison* and *Bos* are closely related to *Bubalus*, as confirmed by morphological, paleontological and reproductive analyses (Arif *et al.,* 2012). According to MacEachern *et al.* (2009), the presence of identical haplotypes between other representatives of the Bovina subtribe and domestic cattle could result from an ancient flow of gënes between these animals before reproductive isolation took place. However, the degree of similarity detected between Yak and domestic cattle across variable, outlier and haplotype sites suggests that hybridisation may have occurred between these species (MacEachern *et al.,* 2009).

1.2.3. Phyletic evolution of the taurine and zebu subspecies (*Bovidae, Bovinae*)

Bos taurus (Linnaeus, 1758) or *Bosprimigenius taurus* is the scientific name given to all the domestic cattle of the ancient world descended from the various subspecies of *Bos primigenius*, the wild Aurochs. The latter was a large animal that once lived in North Africa and Eurasia, from the Atlantic to the Pacific coast. However, it has been extinct in Europe since 1626 and subsequently in other continents. There are two main subspecies: *Bos taurus* (taurines) and *Bos indicus* (zebus). So for a zoologist, talking about cattle implies talking about a being belonging to the animal kingdom, to the Chordes phylum, to the Vertebras subphylum, to the Mammifëres class, a l'ordre des Artiodactyles, a la famille des *Bovidae*, a la sous-famille des *Bovinae*, au genre *Bos*, a l'espëce *taurus* et aux sous-espëces *Bos taurus* et *Bos taurus indicus* (Linnaeus, 1758).

Analysis of mtDNA sequence polymorphism and microsatellites (Loftus *et al.,* 1994; MacHugh *et al.,* 1997) have revealed that the ancestors of taurines and zebus diverged about hundreds of thousands of years ago and must therefore have resulted from at least two biologically independent domestication events. In other words, estimated divergence times indicate a separation of the two lines of cattle of around 2.0 ± 0.14 million years (based on nucleotide sequences) and 1.7 ± 0.50 million years (based on amino acid sequences). These two types of cattle are currently classified as sub-species and differed primitively in the presence or absence of a hump. Moreover, these two subspecies hybridise easily, producing fertile offspring in their entirety (Hiendleder *et al.,* 2008). Analysis of the complete mitochondrial DNA of these two subspecies shows that the sequence of *Bos taurus* A (old reference) determined by Anderson *et al.* (1982) differs from that of *Bos taurus* H (new reference) by 11 nucleotide positions and that of *Bos indicus* H (new reference) by 237 nucleotides (Hiendleder *et al.,* 2008). The number of non-synonymous substitutions in the latter is similar to that observed in interspecific comparisons of mouse and human mtDNA, and 7

which represent focal points in their gënëalogy. Given these occurrences, the hypothesis that the differences observed at the level of the дёпоте mitochondrial, between *Bos taurus* and *Bos indicus* would reflect those observed on their phënotype, is supported (Hiendleder *et al.,* 2008).

1.3. Genetic characterisation methods and markers

Historically, the first efforts to characterise animal populations focused on describing phenotypic variability (Sokouri *et al.,* 2007). Over the last few decades, however, tools for investigating polymorphism in animal resources have become much more diverse, in particular highlighting polymorphism at the DNA level. The use of several types of molecular marker, combined with increasingly sophisticated methodological approaches, has made this genetic information more accessible within different organisms.

1.3.1. Phenotyping methods

The phenotypic characterisation of animal genetic resources uses survey forms as investigation tools, enabling an inventory of the different types of cattle, their qualitative and quantitative description as well as a description of their production environments.

1.3.1.1. Auctions

The surveys are carried out in order to systematically collect the necessary data on the animals and their production environment, thus making it possible to identify and discriminate between cattle breeds according to the conditions in which they are reared. The set of data collected on animal breeds generally concerns: geographical distribution, phenotypic observations, genetic production and reproduction traits, animal use and the characteristics of production environments (Traore, 2010). The animal data collected provide several categories of variables, including basic information, phenotypic descriptors and traits of economic importance. Most of this information is stored in national, regional or global databases (Traore, 2010).

1.3.1.2. Phenotypic description

The description of animal breeds uses descriptors defined by the FAO (2011) for the phenotypic characterisation of cattle. This method is widely used in morphological characterisation studies of animal zoo genetic resources (Sokouri *et al.,* 2007; N'Goran *et al.,* 2008; Coyral-Castel *et al.,* 2009). The most commonly described phenotypic observations are: presence or absence of thoracic hump and its size and shape; coat, skin, muzzle, eyelid, horn and hoof colours; head profile; presence or absence of horns; horn orientation; shape of eyes, horns and ears; ear carriage; dewlap development, tail size; pigment patterns. Added to this are the sex, body condition score and age of the animal, etc.

Different descriptors of dairy aptitude are used to assess the performance of cows in ëlevage. According to N'Goran *et al* (2008), these include: average milk production per day, calving-calving interval, lactation length, age at first calving and number of lactations. According to Banik and Gandhi (2010), the best way to assess genetic value in breeding is to evaluate the genetic traits transmitted by female ancestors to their progeny. Consequently, the evaluation of reproductive and production performance uses the General Linear Model (GLM) procedure to estimate the effects of various genetic and environmental sources on the variation of traits of economic importance (Flores *et al.,* 2007; Khodaei Motlagh *et al.,* 2013). This generalized linear model is used to analyse the effects of independent variables on dependent variables:

$Yij = \mu + Fi + Eij$

Yij = dependent variables (daily milk production, calving-calving interval, length of lactation, age at first calving, etc.) ;

μ = population average ;

Fi = effect of the ith fixed independent variable (year of birth, year and month of calving, age at calving, calving season, calving parity, lactation season, stage of lactation, etc.); Eij = random error associated with observation Yij.

These models consider the following parameters as fixed environmental factors: year of birth, year and month of calving, calving season, age at calving, calving parity, lactation season, stage of lactation; and as economically important genetic traits: milk production per day, lactation duration, age at first calving, calving-calving interval (Morammazi *et al.,* 2007; Banik and Gandhi, 2010; Khodaei Motlagh *et al.,* 2013).

1.3.2. Biochemical markers: allozymes

The first evidence of biochemical variations was provided at the beginning of the 20eme century on human ABO blood groups. However, it was not until the 1960s that the first studies aimed at understanding evolutionary processes using biochemical techniques were carried out (Lewontin and Hubby, 1966).

Thanks to the gel electrophoresis technique, it has been possible to identify protëic variants, and these studies have since become the tool of choice for investigating biochemical variation and providing the first non-biased means of estimating genome variability. By studying 13 allozymes (including 11 blood group loci), Grosclaude *et al* (1990) attempted to clarify the genetic relationships between 18 French cattle breeds. This study led the authors to distinguish four (04) sub-sets of breeds consistent with historical and geographical data. Similarly, Randi *et al* (1991) used the same technique to establish the evolutionary relationships between different species of the genera *Capra*, *Ovis* and *Rupicapra* (Artiodactylus, *Bovidae*).

The first studies of this type in Burkina Faso date back to the work of Queval and Petit (1982) on the polymorphism of hemoglobin in trypanosensitive and trypanotolerant breeds of cattle and their cross-breeds.

Like morphological markers, the major limitations of allozymes are the small number of loci likely to be revealed and the fact that there is a certain organ specificity: not all enzymes are present or active in all organs (Rege, 1992).

1.3.3. Molecular methods for genetic characterisation I.3.3.1. RFLP

Restriction Fragment Length Polymorphism (RFLP) is a molecular biology technique developed by Grodzicker *et al* in 1974. The principle of RFLP is based on the size polymorphism of DNA fragments linked to molecular mutations (nucleotide substitutions, insertion or deletion of nucleotides, rearrangement of nucleotide sequences) produced at the enzymatic restriction sites of the DNA fragment of individuals of a species. In practice, after extraction, the DNA is subjected to one or more restriction enzymes or endonucleases (Meselson and Yuan 1968; Arber and Linn 1969) which cut the double-stranded DNA molecule at specific sites defined by a sequence of bases called restriction sites. These restriction enzymes recognise a very short DNA sequence of 4 to 6 nucleotides, generally double-stranded and usually palindromic (identical in both reading directions). Any mutation in the restriction site sequence prevents the enzyme from acting. This non-cutting of the DNA is detected by a variation in the number and length of the DNA fragments (restriction fragments) obtained after enzymatic digestion, then separation by electrophoresis and visualisation by hybridisation with a radioactive or fluorescent probe. The major advantage of this technique is that it provides easy access to certain nucleotide substitutions, which often represent a high degree of molecular variation. The number of alleles will change as a function of nucleotide mutations in the gene fragments. In animal genetics, the use of this method has made it possible to draw up genetic maps in hens (Bitgood and Somes, 1993), pigs and cattle (Gellin and Grosclaude, 1991), and to identify regions of the genome involved in the variability of quantitative traits, also known as "QTL" (Quantitative Trait loci) in cattle

(Barendse *et al,* 1997), hens and sheep (Pinard-van der laan, 2000).

Since its development, RFLP has been widely used to characterise animal populations. However, this technique became almost obsolete with the advent of technologies such as polymerase chain reaction (PCR), which revolutionised the visualisation of polymorphism in molecular markers, leading to the emergence of other techniques, including PCR-RFLP.

1.3.3.2. PCR-RFLP

PCR is used to amplify a defined region of the genome and then apply the RFLP technique to the PCR product. This method, known as PCR-RFLP, makes it easy to obtain codominant markers and avoids the hybridization step and the use of radioactive probes. The PCR product digested by one (or more) restriction enzymes is simply run in an agarose gel and the polymorphism in the position and number of bands is visualised by a colour reaction with ethidium bromide.

This method has been used to study the level of diversity in populations of wild mammals (Wolf et *al., 1999),* molluscs (Fernandez-Tajes and Mendez , 2007), and domestic mammals (Han et *al.,*2013).

The major drawback of RFLP techniques in general is the possible existence of mutations at restriction sites, which could lead to false results due to the loss or gain of restriction fragments. Furthermore, the very principle of RFLP is based on the use of a very small quantity of genetic material, which means that several enzymes need to be used to correctly identify the subjects under study. The use of such a large number of enzymes can lead to complex matrices that are difficult to interpret. To overcome these drawbacks, a method based on the selection of a limited number of restriction enzymes has been developed: AFLP.

1.3.3.3. AFLP

This technique was first described by Vos and colleagues in 1995. It is based on the selective amplification of restriction fragments. It combines several ësteps including enzymatic digestion and two PCRs. In the first stage, gënomic DNA is cleaved by two restriction enzymes, then adapters of known sëquences and spëcific to the restriction enzymes used are added to the ends of the fragments. A first so-called prësëlective PCR is performed on the fragments obtained using primers spëcific for the adapters. A second PCR (sëlective amplification) uses primers identical to the first, but extended at the 3' end by one to three nucteotides; these sëlective primers reduce the number of amplified fragments to around one hundred. The final product contains 50 to 100 amplified fragments, which can be visualised on a denaturing acrylamide gel by radioactivity or fluorescence. It is possible to observe polymorphism of the presence/absence type of bands on the prints gënërëes. This technique makes it possible to detect several polymorphic sites by PCR reaction, each site corresponding to a unique locus. It is powerful, stable, rapid, and requires no prior knowledge of sëquences of the ëtudië gënome (Najimi *et al.*, 2003). Despite all these qualities, the use of this technique in animal genetics remains very low compared with that of botanical taxa. Over a retrospective period of nine (09) years (from 1995 to 2003), Benshc and Akesson (2005) identified only 115 studies on mammals, birds and fish, but only 33% of these studies were on domestic mammals, compared with 77% on plants over the same period. The major drawback remains the fact that the markers generated by AFLP are dominant; it is therefore difficult to differentiate hëtërozygous individuals from homozygous individuals. This reduces their power when analysing population gënëtics on intraracial and inbred diversity. However, AFLP profiles are highly informative in the evaluation of relationships between breeds (Ajmone-Marsan *et al.*, 2002, De Marchi *etal.,* 2006) and related species (Ajmone-Marsan *etal.,* 2002) as well as in the elaboration of genetic maps (Van Haeringen *et al.,* 2001). Among other things, this technique has been used in studies of the diffërentiation of 51 European cattle breeds (Negrini *et al.,* 2006), in Belgium in the caraerisation of shrimp populations and in the establishment of a partial genetic map of rabbits (Van-Haeringen *et al.,* 2001).

The AFLP technique has proved to be a very valuable tool for studying the genetic structures of animal populations and their diversity. Its high sensitivity provides more information on variability than any other molecular technique. However, it is very expensive in terms of consumables and laboratory reagents, and requires the installation of cutting-edge analysis equipment that is still not accessible to researchers for the interpretation of results. Researchers are now turning to new methods.

1.3.3.4. RAPD

This method consists of PCR amplification of gënomic DNA fragments using arbitrary short (10 bp) primers (Martin *et al.*, 1990). These short, unspecific primers bind randomly to homologous sëquences found in the gënome. Because of its small size, the primer has a high probability of hybridising to complementary sites close to each other and in reverse orientations on the template DNA. The DNA targeted by RAPD is essentially nuclear DNA and in particular rëpëtiactive regions (rëpëtëes). For a given gënome, there may be a number of hybridization sites for this primer. Within a population, mutations influence primer binding sites. At the end of RAPD, a multi locus profile is obtained. The amplification products, visualised on agarose gel in the presence of ethidium bromide, vary in length and in the nature of their sequence.

This technique is recognised as being simple, inexpensive, rapid and does not require any prior knowledge of the sequence. This has justified its widespread use in studies to characterise animal populations. According to Cushwa *et al* (1996), RAPD is the best molecular biology technique for studying the genetic variability of animal populations. In particular, this technique has been used to characterise zebus in Tanzania (Gwakiska *et al.,* 1994), where it was possible to demonstrate introgression of N'dama taurine populations by zebus. RAPD has revealed the existence of a high degree of homogeneity in goat populations (Kantanen *et al.,* 1995). Kumari *et al.* in 2013 used this method to study the genetic revolution of Black Bengal goat populations. In Brazil, RAPD has been used to study genetic variability in 3 breeds of horse (Egito *et al.,* 2007). In Egypt, the genetic polymorphism of 5 camel breeds (Baladi, Somali, Sudani, Maghrabi and Mowallad) has been established (Mahrous *et al.,* 2011). Finally, this technique is also widely used in the production of genetic maps (Martin et *al.,* 1991).

However, RAPD is a technique that lacks reproducibility, as it is highly sensitive to DNA concentration and amplification conditions (Najimi *et al.*, 2003), unlike techniques using microsatellite markers.

I.3.4. Molecular markers

1.3.4.1. Nuclear DNA markers

1.3.4.1.1. Microsatellite markers

Microsatellites or STRs ("Single Tandem repeats") are DNA sequences consisting of tandem repeats (10 to 20 times on average) of a motif of 2 to 6 base pairs, generally no larger than 200 base pairs. They can be perfect (TGTGTGTGTGTGTGTG), interrupted (TGTGTGCATGTGTGCATGTG), or composës (TGTGTGTGTGCACACACACA). These sëquences, ëgalement appelé simple sëquence repeats (SSR), ou variable number tandem repeats (VNTR), are very abundant and well distributed in the eukaryotic gënome (Chambers et MacAvoy, 2000). Microsatellites are predominantly located between genes, in gene introns and in untranscribed regions. In most mammals, (TG)n-type dinucleotides predominate, occurring every 50 to 100 kilobases (Vaiman, 2000). The frequency of microsatellites varies from species to species; every 47 kb in pigs (Wintero *et al.,* 1992) and every 120-180 kb in cows (Steffen *et al.,* 1993). Each microsatellite corresponds to a unique locus in the genome, perfectly defined by the unique sequences that flank the repeat. The length of these sequences varies from one individual to another and from one allele to another within the same individual (Boichard *et al.*, 1998). As the number of repetitions of the motif is variable, it generates a polymorphism in the length of the amplified marker. It is this length polymorphism that is detected during genotyping.

There are several advantages to using microsatellites as genetic markers (Weber and Wong, 1993): (1) the degree of polymorphism due to the fact that the number of repeats can be highly variable in relation to their high rate of molecular evolution; (2) codominant : the heterozygous individual simultaneously displays the characteristics of both homozygous parents and can be distinguished from each of the parents; (3) they are distributed more or less evenly over the entire nuclear genome; (4) these markers are neutral with regard to the selection process and their transmission follows a Mendelian pattern; (5) the different microsatellite alleles are easily identified by simple PCR amplification. All these genetic and technical characteristics make microsatellites the markers of choice for characterisation studies. The FAO recommends the use of microsatellite markers in genetic resource characterisation studies and has drawn up a common list of microsatellite markers for domestic animals. They are currently the most widely used markers in farm animal genetic characterisation studies (FAO, 2011).

1.3.4.2. SNP markers

SNPs (Single Nucleotide Polymorphism) are used as an alternative to microsatellites in studies of genetic diversity. SNPs are the most abundant form of genetic variation in the human genome. They account for more than 90% of all differences between individuals. It is a type of DNA polymorphism in which two chromosomes differ on a given segment by a single base pair. As biallelic markers, SNPs have relatively low quantiles of information and, to reach the level of information of a standard panel of 30 microsatellite loci, larger quantities must be used. In addition, ever-evolving molecular technologies are increasing the automation and reducing the cost of SNP typing. SNPs are likely to be interesting markers to apply in genetic diversity studies in the future, because they can be easily used to assess functional or neutral variation.

1.3.4.3. Mitochondrial DNA markers

Mitochondrial DNA (mtDNA) polymorphisms have been widely used in analyses of phylogenetic and genetic diversity. Mitochondrial DNA has a maternal mode of inheritance (animals inherit mtDNA from their mothers, not their fathers) and a high mutation rate; it does not recombine. These characteristics enable biologists to reconstruct intra- and inter-racial evolutionary relationships by evaluating mtDNA mutation patterns. mtDNA markers can also provide a rapid means of detecting hybridisation between species and subspecies of livestock (Nijman et al., 2003). Polymorphisms in the sequence of the hypervariable region of the D-loop or control region of mtDNA have contributed greatly to the identification of the wild ancestry of domestic species, to the establishment of geographical models of genetic diversity and to the understanding of the domestication of farm animals (Bruford *et al.,* 2003).

1.4. General information on cattle rearing in Burkina Faso

With an annual growth rate of 2%, Burkina Faso's cattle population was estimated at 9.84 million head in 2014 (INSD, 2019). The Sahel, Hauts Bassins, Est and Boucle du Mouhoun regions alone account for 57.35% of cattle numbers (INSD, 2019). As in other West African countries, cattle farming is organised around two (2) sub-species, namely taurins *(Bos taurus)* and zebus *(Bos indicus),* all descended from a common ancestor, the auroch, which is now extinct and originated in the Near East (Lhoste et *al.,* 1991). In addition to these two (2) subspecies, there are the more or less stabilised products of their crossbreeding, known as metis or hybrids, and imported breeds. Given the paucity of genetic characterisation work carried out on local breeds in Burkina Faso, we will limit future descriptions of local cattle to the types that can be found in Burkina. In most African countries, animal breeds are defined on the basis of their membership of a given ethnic group or agro ëcological zone without any prior genetic study; this can lead to a given breed having several different names depending on its geographical location.

1.4.1. Local breeds

1.4.1.1. Sudanese Zebu

The Sudanese Fulani zebu makes up the bulk of the Ь^ИмЬё livestock in the Sahel (MRA, 2003). It is a zebu with lyre-shaped horns (figure 1). It is represented in the subregion by several varieties, including the Djielgobe zebu in Burkina Faso (Larrat, 1988). Its range, which used to be confined to the northern zone (Sahelian and Sudano-Sahelian), has now been considerably extended as a result of successive droughts, which have caused it to descend further and further south. Morphometric descriptions show the Sudanese zebu to be a medium-sized animal, with a long, but not very deep body. The height at the withers is between 1.20m and 1.40m (Benefice *et al.,* 1993). The head is long and thin. The muzzle is dark and the gaskets have many folds. The horns vary in size, but are generally fairly long. The back is plunging and the croup sloping. The chest is deep, but lacks thickness. The legs are long in relation to the body. The well-developed, muscular hump is more prominent in males than in females. The dewlap is thin but very wrinkled, extending from the chin to the rear. The coat is smooth and short. The coat is generally grey or light grey with dark patches. The dominant coat is light grey with mucous membranes that are often black in some herds (Benefice *et al.,* 1993). The skin is soft, with pigmentation varying from light to dark. The udder and teats are poorly developed. The average birth weight of calves is about 17 kg for males and about 15 kg for females. At adulthood, live weight varies between 300 kg and 350 kg for males and 250 kg to 300 kg for females (Nianogo *et al.*, 1996). The milk production of the Sudanese Peul zebu is modest, but its milk is rich in fat (5.5%) (Larrat, 1988). Total production is 700 kg in 8 months of lactation (Benefice *et al.,* 1993). A study estimating the additive genetic value of Peul zebu showed a strong correlation (0.57) between height at withers and milk production. However, the correlations are negative between conformation traits (body depth or musculature, breeches) and milk production performance (Sanoga, 2003). This situation indicates that any selection in favour of breeching and muscling will be accompanied by a dëfavoraЬle evolution of milk production. A dual-purpose sëlection, i.e. for dairy and beef production, would be completely inefficient, which justifies the extensive use of this animal for meat production. It is in fact a good meat animal, with an average weight of 300 kg and a meat yield of 48%. Top-quality carcasses reach 50% and are exported in large quantities (Benefice *et al.,* 1993).

Figure 1: Sudanese Zebu peul (Photo Tapsoba, 2017)

1.4.1.2. Baoule bullfighting

The Lobi country (in the south-west) concentrates the majority of Baoule bull populations in Burkina Faso (Ouedraogo, 1989). Trypanotoierente is the most widespread taurine in Burkina Faso; originating in south-west Burkina in the Lobi country, it is also found in the Baoule country in Côte d'Ivoire (hence its name). The first descriptive studies date back to 1947 when Doutressoulle described Baoule taurines as small, straight, short and ellipometric cattle. The height at the withers varies from 90 centimetres to 1 metre. The head is broad and short, with a straight muzzle. The well-marked orbital arches give the broad forehead a certain concavity. The horns are short, broad, horizontal and circular in the male and oval in the female. There are often individuals without horns or with floating horns. The dorsal line is more or less straight from the withers to the base of the tail and the back is broad and well muscled. The trunk is rounded, but rather strapped behind the shoulder. The hindquarters, of medium length, are slightly sloping and the spine is not very pronounced. The long tail ends in a full tuft. The dewlap and ventral fold are not very noticeable. The limbs are short and slender. The extremities are dark, black or marked with black. The coat is black piebald, black piebald, yellow piebald or yellow, rarely fawn or wheaten. The extremities are dark, black or marked with black (Figure 2).

The average birth weight is 22.7 kg. The average weight is 190 kg for cows, 200 kg for bulls and 220 kg for steers. Despite its small size, the Baoule bull's zootechnical performance is interesting, which justifies its dual use in meat and milk production.

Figure 2: Taurin Baoule (Photo Tapsoba, 2017)

1.4.1.3. Mere

The Mere is a cross between the Fulani zebu and the Baoule taurine. The birth of this product can be attributed to the migratory movements of Fulani populations in search of pasture. The Mere also meets the need to improve production because of the relatively low productivity of Baule cattle. This is made possible by the advance of the desertification front, which has made areas with a high prevalence of tsetse fly, such as the south, south-west, centre-west and centre-south of Burkina Faso, accessible to the zebu peul. Morphologically, the Mere is closer to the Sudanese zebu peul, whose hump it bears after several generations of cross-breeding.

1.4.2. Imported breeds

1.4.2.1. Zebu M'bororo

Mason (1951) classifies the M'bororo clans in the category of zebu with high lyre horns, to distinguish them from Fulani zebu with lyre horns. Also known as Fulani Blanc, Foule Blanc or Akou, the M'bororo zebu is bred by the Fulani in the central Sahel region. With its fine bone structure, the M'bororo zebu is a large, tall-legged animal with long, slender limbs and strong hooves. The musculo-fatty hump, which is fairly well developed, is located in the cervico-thoracic region. It is much larger in bulls than in cows. The dewlap, also developed, extends from the chin to the sternum. The ventral fold is mobile and hanging, as is the male's sheath. The head is long and slender, with large, high, open, erect lyre-shaped horns, generally white in colour, measuring between 75 and 120 centimetres. The back is long, but the side lacks roundness and appears flat due to the narrowness of the shoulders. The bittern is sloping. The skin is loose and of medium thickness, with pigmentation varying from light to dark. The coat is short and coarse. Coat colour varies from reddish brown to tan and the tuft is sometimes white (Figure 3). The M'bororo zebu is fierce, shy and difficult to train.

The cows are very poor milkers, producing around 2 litres per day at peak times. According to Salifou *et al* (2012), the slaughter yield is 61.1%. The skins are highly prized and yield good quality hides. These animals have a reputation for obeying their masters' commands like dogs, making them good 'bush' animals (Shaw and Colville, 1950). The M'bororo is a very good walker, perfectly suited to transhumance herding.

Figure 3: Bororo Zebu (Maaouia, 2018)

1.4.2.2. Zebu Azawak

The Azawak zebu takes its name from its area of expansion called AZAWAGH in Tamashek, i.e. northern or sandy country with no marked relief (Seydou, 1981) (Figure 4). Burkina Faso imported Azawak cattle in 1967 and 1969 for the Markoye station, and these cattle were subsequently sold to breeders in the Sahel and other parts of the country (Lakouetene, 1999). It is one of the best milkers in the Sahel, producing an average of 3 to 4 litres a day on meagre pasture. It is a good slaughter animal, producing 500-600 kg after 5-6 years of fattening; its carcass yield is 50% (Oumarou, 2004).
In addition to these two breeds, there are many other imported breeds such as the Arab zëbu, Goudali and Djeli.

Figure 4: Zebu Azawak (Maaouia, 2017)

1.5. Cattle breeding systems

As in other countries in the sub-region, there are two main types of livestock farming system in Burkina Faso (MRA, 2012[a]). This classification, based on agro-ecological potential (Sere and Steinfeld, 1996) and correlated with socio-cultural and economic factors, makes it possible to distinguish between so-called traditional and improved systems:
-Traditional systems: Animal housing is precarious and often non-existent. Health protection is limited to compulsory periodic vaccination campaigns initiated by the state. Livestock production is essentially supported by natural grassland and shrub resources, which provide a virtually free food supplement for animals grazing mainly on uncultivated land. The

availability of grazing land dictates the movements of the herds, which define the nomadic or extensive sedentary and transhumant modes of production.
The transhumant pastoral system is widespread throughout the country, with a strong preponderance in the Sahel, eastern and western regions (in the cotton basin):

✓ the large-scale transhumance system in which herds move over great distances. These distances are often, but not exclusively, cross-border, which implies a change of agro-ecological zone (MRA, 2005).

✓ the small-scale transhumance system, in which the animals' movements, spread over short periods, are confined to neighbouring localities. These movements are seasonal and allow access to pastures, especially during the rainy season. Once this period has passed, the animals return to their base.

In traditional systems in general and in the transhumant pastoral system in particular, food supplements are almost non-existent, there is no crop-livestock integration, the building up of fodder reserves and fodder crops are low. However, the use of post-harvest fodder is an important link in the grazing chain. This complementary link between crops and livestock is, however, limited to the use of crop residues through grazing. Water availability is difficult in the dry season; animals are watered from surface water and sometimes from the few boreholes that exist in the grazing areas crossed (MRA, 2012).[b]

Extensive sëdentary farming systems: in these systems, the distance travelled by the animals when on the move does not exceed one day's walk from the farm or the night pen. In this system, the owners practise a lot of confiage.

-Improved livestock production systems: These systems are generally being developed in towns and outlying areas, especially around major cities such as Ouagadougou. Their growth is helping to meet a growing demand for animal protein that cannot be met by the long supply and marketing networks of small-scale farmers (OECD, 2008). These systems are semi-intensive or intensive. They require more substantial investment in infrastructure, inputs and labour. The animals are vaccinated and their health is monitored regularly. These types of farming are the initiative of civil servants, traders, retired people, young farmers and political decision-makers who invest for commercial purposes (MRA, 2010). The feed complement is made up of agro-industrial by-products (SPAI) and industrial feeds such as concentrates (MRA, 2012[a]). Despite these investments, these farms are unable to meet market requirements either quantitatively or qualitatively (OECD, 2008). This situation can be attributed to the various constraints affecting the livestock sector in Burkina Faso.

CHAPTER II

MATERIALS AND METHODS

11.1. Morpho-biometric characterisation

11.1.1. Description of the study area

This study was carried out in the Centre-West, Centre-South, South-West and Sahel regions of Burkina Faso.

11.1.1.1. Centre West Region

The Centre-West region covers an area of 21,752.48 km^2 , or 8% of the national territory. It is bordered to the east by the Centre-Sud and Centre regions, to the north by the Nord region, to the west by the Boucle du Mouhoun and South-West regions and to the south by the Republic of Ghana. It comprises four provinces (Boulkiemde, Sanguie, Sissili, Ziro), divided into 39 communes. The capital of the region is Koudougou.

The region is characterised by a flat topography with some elevations, especially in the Sanguie province. Depending on the province, there are sandy-clay soils, ferruginous soils, thick, loose ferrallitic soils and pockets of hydromorphic soils in the low-lying areas.

The climate is Sudanian, with average annual rainfall of between 700 and 1000 mm, rising to 1200 mm in the provinces of Sissili and Ziro (Guinko, 1984). The lowest temperatures (on average 12°C) are generally observed during the months of December and January, while the highest (on average 38°C) occur between March and May. The region is characterised by three types of vegetation. From north to south, there is a shrubby savannah typical of the northern Sudanian climate, a shrubby savannah typical of the southern Sudanian climate, and gallery or open forests found mainly along watercourses.

The population of the Centre-West region is estimated at 1,510,975 (INSD, 2016). The Gourounsis and Mossis are the two main ethnic groups in the Centre West region.

The two main economic activities in this central-western region are agriculture and livestock farming. Sedentary livestock farming is practised by farmers, who usually keep cattle, small ruminants, donkeys, pigs, horses and poultry. Transhumant herding is practised by the Peuhl, whose herds are made up of zebus and metis zebus x taurins. The region's cattle population is estimated at 706,000 head (MRA, 2014). Bulls are mainly bred by the Gourounsi ethnic group, hence the name "Gourounsi".

taurin Gourounsi. Trypanosomosis is a limiting factor for elev;ige in this region. At least 60,000 cattle are milked each аппëе against trypanosomosis (MRA, 2014).

11.1.1.2. Centre-Sud Region

It is located in the south of Burkina Faso, covering an area of 11,327 km^2 , or 4.1% of the national territory. It is bordered to the south by the Republic of Ghana, to the east by the Centre-Est and Plateau Central regions, to the north by the Centre region and to the west by the Centre-Ouest region. The region is divided into three (3) provinces (Bazega, Nahouri and Zoundweogo) with 19 communes and 528 villages. The capital of the region is Manga.

There are two main topographical areas in the Centre-Sud region: plains and plateaux. The soil is made up of large hydrogeological units unsuitable for crops, with raw mineral soils suitable for agro-pastoral activities.

The climate is Sudano-Sahelian, with rainfall ranging from 700 mm to 1,200 mm (Guinko, 1984). The average monthly temperature is around 30°C. The vegetation is essentially composed of open shrub savannah, dense shrub savannah, degraded shrub savannah, dense shrub savannah and gallery forests along watercourses. The hydrographic network is essentially made up of the Nakambe, Nazinon and Sissili basins, with numerous periodic tributaries.

The population of the Centre-Sud region was estimated at 804,709 in 2015 (INSD, 2015). Three major indigenous ethnic groups dominate the region: the Mossi, the Gourounsi and the Bissa. The region's economy is based mainly on agriculture, livestock farming, fishing and trade. The cattle population is estimated at 318,000 head (zebus and bulls), with an annual growth rate of 2% (MRA, 2014). Gourounsi cattle are mainly bulls, which they use as dowries at customary weddings. The region also suffers from tsetse fly pressure, which means that zebu farmers in the area have to be regularly treated. Bulls are rarely treated.

11.1.1.3. South West Region

The South-West region is bordered to the east by the Republic of Ghana and the Centre-West region, to the south by the Republic of Côte d'Ivoire, to the west by the Cascades and Hauts Bassins regions and to the north by the Boucle du Mouhoun and Centre-West regions. Its total surface area is approximately 16,576 km^2 or 6.1% of the national territory. It is 24

Composed of four (4) provinces (Bougouriba, Loba, Noumbiel and Poni) organised into 28 communes. The capital of the region is Gaoua.

The South-West is one of Burkina Faso's best-watered regions (between 900 and 1200 mm isohyëtes). It belongs to the Sudano-Guinean climatic zone and is characterised by a very uneven relief, an average annual tempërature of 27°C which fluctuates between a minimum of 21°C and a maximum of 32°C. The vegetation is relatively dense and varied. From north to south, there is an evolution from wooded savannah to open forest and forest galleries along the watercourses. The river system is part of the Mouhoun river basin, with a large number of potential dam sites (Bougouriba, Noumbiel).

The population of this south-western region is estimated at 795,549 (INSD; 2016). The main ethnic groups are : Lobi, Dagara, Gan, Birifor. The population is predominantly rural and agricultural, with animals used mainly for marriage exchanges and customary practices. There are 343,000 head of cattle, mainly Lobi/Baoule bulls (MRA, 2014). Livestock farming is essentially family-based and sedentary, and the animals rarely receive supplementary feed or veterinary care. Trypanosomiasis is one of the main diseases hampering the development of livestock farming in this area (MRA, 2000). However, livestock farming is a promising activity given the potential it offers. These include the wide availability of grazing land and the existence of water points.

11.1.1.4. Sahel Region

The Sahelian zone extends across the north of the country between latitudes 13°5'N and 15°3'N. It is characterised by low rainfall (less than 600 mm) and high temperatures ranging from 15°C to 47°C. The natural plant formations are savannahs and steppes of thorny shrubs, tiger scrub, gallery forests and lowland formations that are sometimes dense (Guinko, 1984). This zone comprises five provinces (05): Oudalan, Seno, Sanmatenga, Gnagna and Yatenga.

11.1.2. Sampling sites

Sampling for the morphological study of Burkina was carried out in the provinces of Sanguie (Centre-West Region), Nahouri (Centre-South Region) and Poni (South-West Region) respectively. The zëbuses that have ële ийНзёз to allow inference of variability have ëГë ëchantШonnës in the provinces of l'Oudalan and Soum in the region of l'Oudalan.

Sahel (Figure 5).

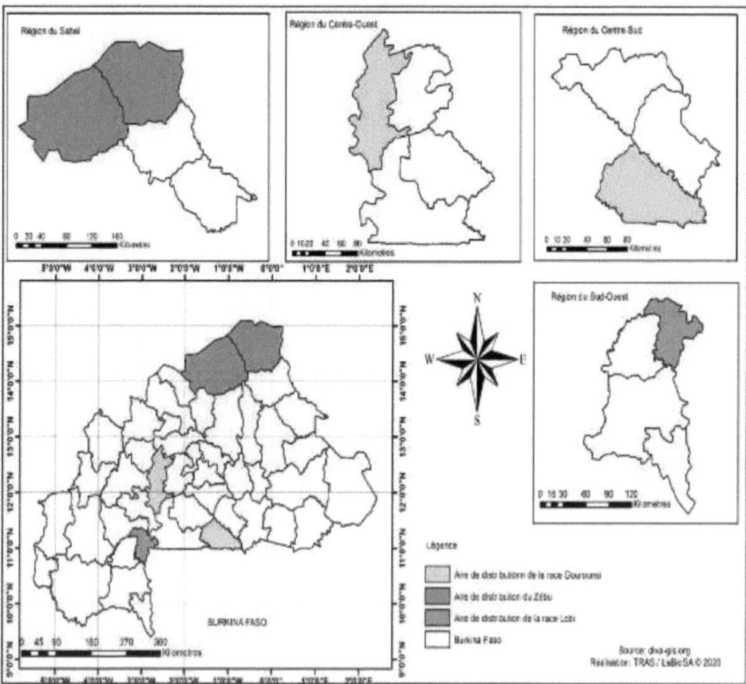

Figure 5: Location map of the study area (Tapsoba, 2020)

II.1.2. Material

11.1.2.1. Organic materials

The study involved cattle of the local bull breeds "Lobi" and "Gourounsi" with a minimum age of 5 years. A total of 1055 cattle were measured, including 264 Lobi bulls (figure 6), 545 Gourounsi bulls (figure 7) and 246 Fulani zebu.

Figure 6: Lobi bulls (Photo Tapsoba, 2017)
Figure 7: Gourounsi bullfighters (Photo Tapsoba, 2017)

11.1.2.2. Measuring equipment

The various body measurements (appendix 1) were ële made using a zootechnical measuring stick measuring approximately 2 metres and an ANImeter ® brand zoometre tape measure (figure 8). Head and body measurements, i.e. muzzle circumference, horn length, ear length, thoracic perimeter, scapulo-ischial length, ischial width and tail length, were taken using the zoometer tape. Measurements relating to height at withers, height at sacrum, depth of chest, width of chest, width at shoulders and width of hip were taken using a zootechnical cane.

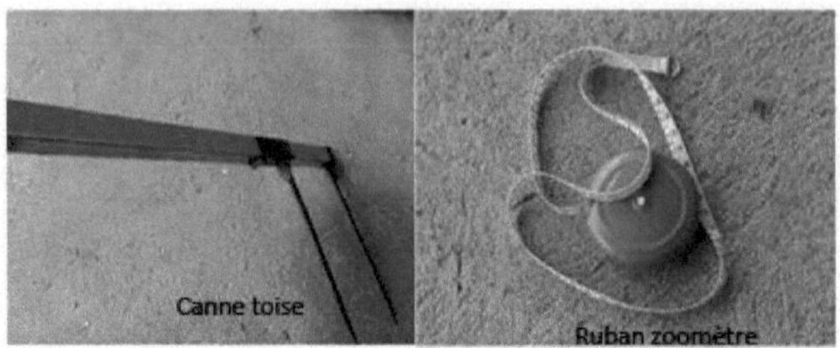

Figure 8: Surveying equipment (*Photo Tapsoba, 2017*)

11.1.3. Methods

11.1.3.1. Field survey methodology

The mëthodology consistedë first of identifying the ëleyeuгз of bullfighters followed by an interview with them. This interview consisted of explaining to them the purpose of this work and the survey procëdures.

11.1.3.2. Sampling

The aim of this sampling is to capitalise on the effect of time on morphological variations in different bull breeds. However, the samples were collected at 4-year intervals in 2014 and 2018. To limit the effect of environmental factors on the body condition scores of the animals sampled, in particular the unavailability of pasture, the samples were all collected at the same time between August and October 2014 and 2018.

In detail, sampling was carried out in five provinces (Sanguië, Nahouri, Poni and Oudalan and Soum) and only involved adult cattle (at least 5 years old). This choice is explained by a tendency to standardise biometric parameters at this age. To avoid kinship between the individuals sampled, a minimum distance of 3 km was observed between herds. The number of herds sampled in each village was a maximum of 3, and the number of individuals retained in each herd was between 1 and 4; however, in herds with more than 80 head, the number of individuals sampled could reach 8.

The total number of animals sampled was 1055, including 820 females and 235 males. The majority of these individuals belonged to the *Bos taurus* genetic type and were distributed as follows: Gourounsi taurines (431 females, 114 males), Lobi taurines (197 females, 67 males). The zebus numbered 246, including 192 cows and 54 bulls. The females were between 5 and 17 years old and the males between 5 and 8 years old. These animals were sampled over two years (with the exception of zebu) in 64 villages in the 5 provinces of the delude zone (Table I). The high number of females compared with males is justified by the fact that they are numerically more important and spend a longer time in the herds, whereas the males are exploited at a relatively young age. In addition, this sampling is geared towards a ratio of "number of females sampled/total number of individuals sampled" of 80%. This ratio is recommended by the FAO for morphological characterisation studies (FAO 2011).

Table I: Sampling description

	2014		2018	
Populations	Workforce	Villages	Workforce	Villages
1.GourN	130	11	128	8
2.GourS	138	10	149	10
3.lobi	125	8	139	13
4.Zebu	246	14	-	-
Total	639	27	416	23

GourN: Goursounsi Nahouri; - GourS: Gourounsi Sanguie

11.1.3.3. Morpho-biometric surveys

The morpho-biomëtric surveys concerned animals ëchantIIIonnës and were carried out in three stages: The data collection methodology was based on the manual for training interviewers in morpho-biometric surveys of local West African cattle breeds as part of the CORAF/Introgression 03.GRN.16 project.

The first stage involved recording information on the age of the animals according to the owner's opinion and the dentition. Other summary animal identification data (survey number, sex, apparent presence, location, breeder) were also recorded.

The second involved measurements of the head, horns and body. A total of 21 measurements were taken on each animal. The animals were kept as still as possible on their feet, without being tensed or strained. Given the survey conditions in the field, most of the measurements were carried out on animals kept in the pens or at their grazing places. In some cases, when facilities allowed, the measurements were taken in the restraint corridors.

❖ **Head and skull measurements**

The quantitative measurements considered at head level relate to 5 characteristics: the length and width of the head, the length and width of the skull and finally the length of the face. The circumference of the snout corresponds to the diameter measured just behind the muzzle. The anatomical points at which these different parameters are measured are shown in Figures 9 and 10.

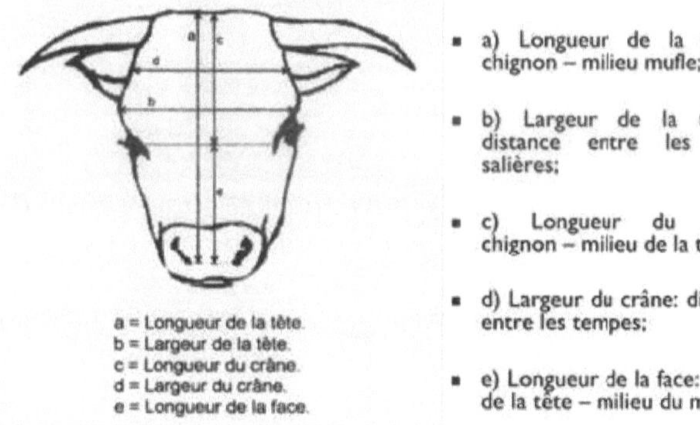

- a) Length of head: bun - middle of nose;
- b) Width : distance between the two salt;
- c) Crane length : chignon - middle of the head;
- d) Width of skull: distance between the temples;
- e) Length of face: middle of head - middle of muzzle.

Figure 9: Anatomical points for head measurements (CORAF/Introgression03.GRN.16)

Figure 10: Measurement of snout circumference (CORAF/Introgression03.GRN.16.)

- **Horn and ear measurements**

The measurements taken on the horns relate to the length of the horns. Horn length measures the distance from the tip to the base of the horn as shown in figure 11. Ear length measures the distance between the base of the conch and the tip of the ear. Anatomical points for horn and ear measurements are shown in Figures 11 and 12.

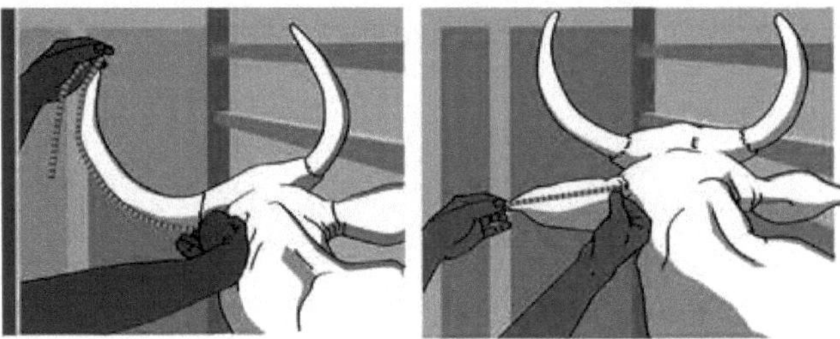

Figure 11: Measurement of horn length Pointe-Base (CORAF/Introgression03.GRN.16.)
Figure 12: Ear length measurement (CORAF/Introgression03.GRN.16.)

- **Body measurements**

The quantitative parameters for body measurements are: height at withers, thoracic perimeter, height at sacrum, scapulo-ischial length, depth of chest, width of chest, width at hips, width at pins, length of pelvis and length of tail.

- **Height at the withers:** this measurement is taken with great care using a zootechnical cane toise, making sure that the animal is on a flat, horizontal surface and that its posture is normal. The measurement is made by holding the measuring stick vertically next to the animal's front limb (right or left), and placing it on the animal's withers, just behind the hump (if present) (figure 13 A).
- **The height at the sacrum:** this measures the vertical distance between the ground and the point determined by the intersection of the line passing through the points of the hips and the sacrum (figure 13 C).
- **Chest depth:** this is measured as the straps pass just behind the front legs (figure 13 B).
- **Shoulder width:** this is the distance between the lateral tuberosities of the two humeri (Figure 13 D).
- **Thoracic perimeter:** this is measured with a tape measure at the shoulder just after the animal's anterior limbs. This measurement is taken just in front of the hump (if present)

(figure 13 F).

- **Scapulo-ischial length:** Scapulo-ischial length is measured using a tape measure and extends from the tip of the shoulder to the tip of the buttocks, the animal ëbeing тшоЬШвё (figure 13 G).
- **Body length:** the length of the body is measured from the point of the shoulder to the wing of the ilium (Figure 13H).
- **Width at the hips**: measured between the two axillae of the ilium (Figure 13 E).
- **Tail length:** measured from the first caudal vertebra to the end of the tail. (Figure 13 I).

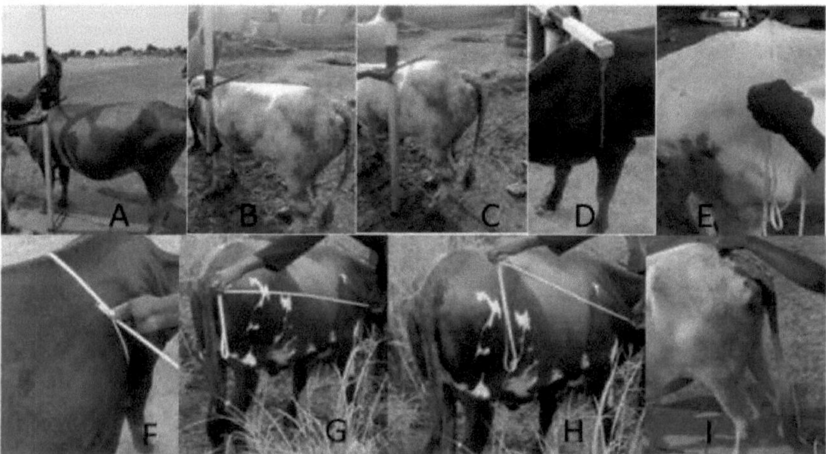

Figure 13: Body measurements (Photo BERE, 2017)

- **Chest width:** this is the distance between the axis of the front limbs and the base of the sternum (Figure 14).

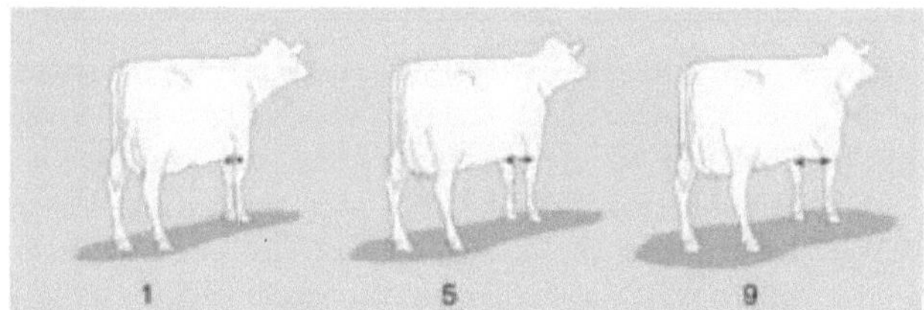

Figure 14: Breast width measurement (CORAF/Introgression03.GRN.16.)

- **Ischial width:** this measures the distance between the two points of the ischium (Figure 15).

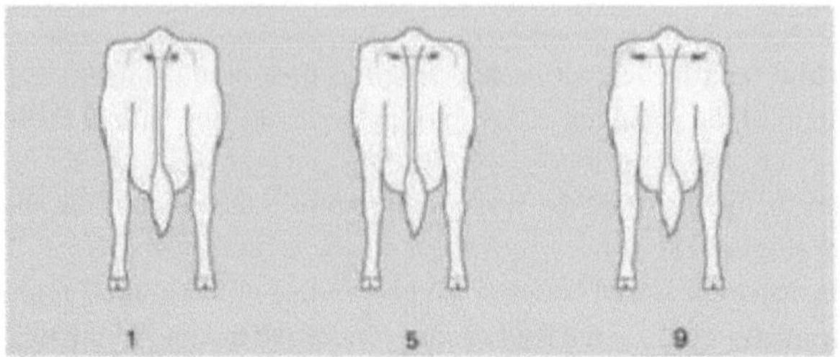

Figure 15: Measurement of Ischial Width (CORAF/Introgression03.GRN.16.)

II.1.4. Statistical analysis

The survey data were ële entered into an Excel spreadsheet in which a preliminary clean-up was ëlë carried out. This involved removing outlier records and deleting variables with more than 30% missing and/or outlier data. The dataset resulting from the first cleaning was transformed into a text file and imported into the R studio software (RStudio team, 2020). The dataset was then normalized using the Zscore and the outliers removed.
Statistical analyses were carried out using R software (R Core Team, 2013). Data collected in 2014 and 2018 were compared with the same measurements obtained on Fulani zebu whose samples were used as an "out group" in order to infërer variability. A principal component analysis was performed using the factoextra package (Kassambara, 2015). After examining the eigenvalues of the different dimensions, variables contributing only very weakly to the formation of the first 2 dimensions were removed in order to amëiorate the total inertia on the first 2 dimensions. These were the variables body length and shoulder width. Then, a bottom-up classification 33
The best possible number of k was determined using the Nbclust package (Charrad *et al*, 2014). The estimation of the level of introgression between the different breeds from 2014 to 2018 was done using linear discriminate analysis (LDA). This was used to estimate the proportion of well-classified animalsës in their original populations. The GLM package was used to evaluate the effect of breed on the various measurements. The modële was refined by including race and year of collection as fixed effects and the race*period interaction The least squares mean of the different measurements and their standard errors were obtained using the emmeans package (Lenth, 2019). To infer the effect of time on the variation in morphological measurements, all the data collected (2014 and 2018) were merged and the measurements on zebus were removed (these measurements were not collected in 2018), then a Multivariate Analysis of Variance (MANOVA) was performed. The model included all the measurements taken as dependent variables. In addition, all variables that changed significantly from one collection period to another were identified by performing a Multivariate Analysis of Variance with race as the independent variable.

II.2 Molecular characterisation of bull breeds in Burkina Faso

11.2.1. Study sites

The study sites correspond to the same localities where the phenotypic surveys were carried out, with the exception of the Sahel region (Figure 16). The choice of these provinces is justified by the fact that they cover the two agro-ecological zones (Sudano-Guinean and Sudano-Sahelian zones) of Burkina Faso in which the bull breeds of Burkina Faso are found.

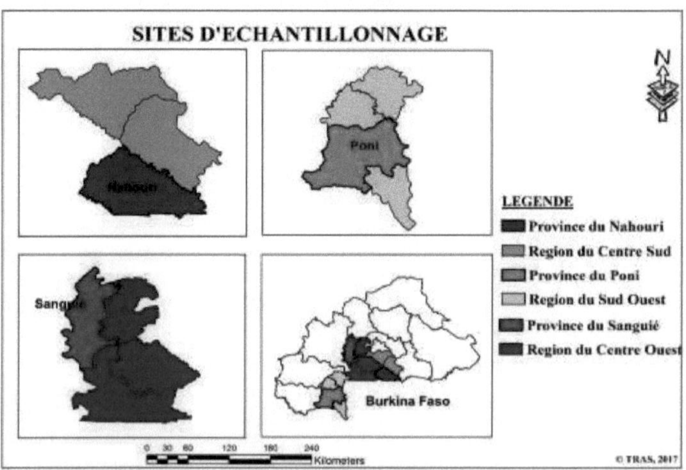

Figure 16: Molecular characterisation sampling sites (Tapsoba, 2017)

11.2.2. Study population

In the context of this molecular caracterisation study, the study population was composed of individuals belonging to the two bullfighting populations (Taurin Lobi and Gourounsi). **II.2.3. Sampling**

This was a random sample in which subjects supposed to be purebred and not apparent were taken into account. A total of 143 bulls were sampled 38 Gourounsi bulls from the Nahouri province, 43 Lobi from the Poni province and 62 Gourounsi from the Sanguie province.

11.2.3. Blood sampling

The biological material consisted of whole blood obtained by puncture of the animal's jugular vein and collected in tubes with EDTA (ethylene diamine-tetra-acëtic acid) anticoagulant. The blood samples were then stored at ± 4°C until extraction of the genomic DNA.

11.2.4. Genomic DNA extraction

DNA, a ël e extracted from whole blood with a purification kit, MasterPure DNA Purification Kit (Biozym Illumina Inc, USA). The extraction ëlë done according to the manufacturer's recommendations. The extracted DNA is stored at 4°C until genetic amplification (PCR).

11.2.5. PCR amplification and genotyping

Conventional PCR has ëlë effecWe for each of the 27 microsatellite markers chosen for this ëtude (Table II). The microsatellites usedës (forward primers: sense) were ëlë cut with one of three fluorescent dyes (FAM, HEX or ATTO550). PCR conditions ëwere as follows: initial denaturation of DNA at 95°C for 15 min, followed by 40 cycles of dënaturation at 95°C for 50 sec, hybridisation of primers for 50 seconds at 53°C, 55°C, 57°C, 58°C, 60°C (depending on the optimal hybridisation temperature for each marker) and a ëlongation at 72°C for 1 min with a final ëlongation of 10 min at 72°C. PCR products were then ëlësubjected to ëlectrophorëse after multiplexing, in an automated DNA analyserë ABI3100 (Applied Biosystems, USA) with ROX500 (Applied Biosystems, USA) as an internal molecular weight marker. The 27 microsatellite loci were multiplexed into five panels for gënotyping, as shown in Table VIII. Allele sizes for each ëample were then ëlë extracted using GeneMapper v.4.1

software (Applied Biosystems, USA).

Table II: List and characteristics of markers used

Multiplex	Locus	Allele size interval	Primer	Fluorescent dyes	Hybridization temperature
1 1	CSRM60-F CSRM60-R	79-115	AAGATGTGATCCAAGAGAGAGGCA AGGACCAGATCGTGAAAGGCATAG	FAM	60°C
1 1	CSSM66-F CSSM66-R	171-209	ACACAAATCCTTTCTGCCAGCTGA AATTTAATGCACTGAGGAGCTTGG	FAM	60°C
1 1	HEL1-F HEL1-R	99-119	CAACAGCTATTTAACAAGGA AGGCTACAGTCCATGGGATT	HEX	56°C
1 1	INRA063-E INRA063- R	167-189	ATTTGCACAAGCTAAATCTAACC AAACCACAGAAATGCTTGGAAG	HEX	56°C
2 2	BM1824-F BM1824-R	176-197	GAGCAAGGTGTTTTTCCAATC CATTCTCCAACTGCTTCCTTG	ATTO550	61°C
2 2	ETH152-F ETH152-R	181-211	TACTCGTAGGGCAGGCTGCCTG GAGACCTCAGGGTTGGTGATCAG	FAM	60°C
2 2	HAUT27-F HAUT27-R	120-158	TTTTATGTTCATTTTTTGACTGG AACTGCTGAAATCTCCATCTTA	HEX	54°C
2 2	INRA05-F INRA05-R	135-149	CAATCTGCATGAAGTATAAATAT CTTCAGGCATACCCTACACC	FAM	54°C
3 3	BM1818-F BM1818-R	248-278	AGCTGGGAATATAACCAAAGG AGTGCTTTCAAGGTCCATGC	HEX	60°C
3 3	ETH3-F ETH3-R	103-133	GAACCTGCCTCTCCTGCATTGG ACTCTGCCTGTGGCCAAGTAGG	FAM	63°C
3 3	HEL9-F HEL9-R	141-173	CCCATTCAGTCTTCAGAGGT CACATCCATGTTCTCACCAC	ATTO550	56°C
3 3	ILSTS006-F ILSTS006-R	277-309	TGTCTGTATTTCTGCTGTGG ACACGGAAGCGATCTAAACG	FAM	54°C
3 3	TGLA53-F TGLA53-R	143-191	GCTTTCAGAAATAGTTTGCATTCA ATCTTCACATGATATTACAGCAGA	HEX	55°C
4 4	HEL5-F HEL5-R	145-171	GCAGGATCACTTGTTAGGGA AGACGTTAGTGTACATTAAC	FAM	54°C
4 4	HAUT24-F HAUT24-R	104-158	CTCTCTGCCTTTGTCCCTGT AATACACTTTAGGAGAAAAATA	HEX	53°C
4 4	SPS115-F SPS115-R	234-258	AAAGTGACACAACAGCTTCTCCAG AACGAGTGTCCTAGTTTGGCTGTG	FAM	61°C
4 4	INRA032-F INRA032- R	160-204	AAACTGTATTCTCTAATAGCTAC GCAAGACATATCTCCATTCCTTT	ATTO550	56°C
5 5	ETH185-F ETH185-R	214-246	TGCATGGACAGAGCAGCCTGGC GCACCCCAACGAAAGCTCCCAG	ATTO550	65°C
5 5	ILSTS05-F ILSTS05-R	176-194	GGAAGCAATGAAATCTATAGCC TGTTCTGTGAGTTTGTAAGC	FAM	56°C
5 5	INRA035-F INRA035- R	100-124	ATCCTTTGCAGCCTCCACATTG TTGTGCTTTATGACACTATCCG	FAM	60°C
5 5	HEL13-F HEL13-R	178-200	TAAGGACTTGAGATAAGGAG CCATCTACCTCCATCTTAAC	HEX	54°C
5 5	TGLA126-F TGLA126-R	115-131	CTAATTTAGAATGAGAGAGGCTTCT TTGGTCTCTATTCTCTGAATATTCC	HEX	54°C
6 6	BM2113-F BM2113-R	122-156	GCTGCCTTCTACCAAATACCC CTTCCTGAGAGAAGCAACACC	FAM	63°C
6 6	ETH10-F ETH10-R	207-231	GTTCAGGACTGGCCCTGCTAACA CCTCCAGCCCACTTTCTCTTCTC	FAM	61°C

6	ETH225-F	131-159	GATCACCTTGCCACTATTTCCT	ATTO550	63°C
6	ETH225-R		ACATGACAGCCAGCTGCTACT		
6	INRA023-F	195-225	GAGTAGAGCTACAAGATAAACTTC	ATTO550	58°C
6	INRA023- R		TAACTACAGGGTGTTAGATGAACTC		
6	TGLA122-F	136-184	CCCTCCTCCAGGTAAATCAGC	HEX	58°C
6	TGLA122-R		AATCACATGGCAAATAAGTACATAC		

11.2.6. Statistical and genetic analysis

❖ Genetic variability

Micro-Checker software version 2.2.3 (Oosterhout *et al.,* 2004) has ële ииНзё for identification and classification of null alleles. A null allele corresponds to the non-amplification of a given allele following a mutation in one of the *primers*. Pairing is no longer made with the complementary sequence at the locus to be amplified - this is referred to as a null allele. By definition, a null allele cannot be detected except in the homozygous state (absence of a band). In the heterozygous state, it will be dominated by the alleles with which it is heterozygous. The presence of these null alleles then leads to a significant deficit in biologically inexplicable heterozygotes (De Meeus, 2012), hence the importance of detecting and eliminating them.

The basic diversity indices have been determined by a set of parameters which are :

> Average number of alleles observed per locus (Na)

This is the average number of alleles observed in a population; it reflects the population's richness in alleles. In general, the average number of alleles depends on the size of the sample because of the presence of unique alleles at low frequencies in the populations, but also because the number of alleles observed increases as the size of the population increases. This parameter is determined by locus and by population. The mean number of alleles per locus (Na) was estimated in the present study using Microsatellite Analyzer (MSA) software version 3.15 (Dieringer and Schlotterer, 2003).

> Polymorphism Information Content (PIC)

PIC is used to assess the informative (discriminative) capacity of a marker in a population based on allelic frequencies (Thiruvenkadan *et al.,* 2014). The PIC is evaluated per locus by considering the total population and per population for all loci combined. This parameter was evaluated using CERVUS software (Kalinowski *et al.,* 2007).

> Observed heterozygosity (HO)

This is the proportion of heterozygous individuals observed at the K locus as indicated by the following formula :

HOK=E(**i#j)aki,j=l**

Where pij is the estimated frequency of genotype ij at locus k and ak is the number of alleles at locus k. If we consider the locus, the rate of heterozygotes observed (HO) is the average of (HOK) according to the equation :

HO= **llEHOKlk=l**

Observed heterozygosity (HO) is measured per locus and per population.

> Expected heterozygosity (HE)

The expected hëtërozygosity (HE) can be calcиlëe under the Hardy- Weinberg equilibrium hypo^se, from the altered frequencies dëterminëes for each locus, using the following formula:

HE= 1- **Epi**^{2}

Or pi is the frequency of the iëme alltie at this locus.

The average rate of hëtërozygosity is the most satisfactory index of gënëtic diversity. Its numëric value dëepends on the number of polymorphic loci and the gënotypic structure of each of them. However, Nei (1978) suggests using an unbiased estimator (HEnb) when the number of animals tested is small. This is defined as ëbeing the probability of drawing, at random, two different al^les at the same locus. The estimate of the unbiased hëtërozygosity is calculated according to the following formula:

HEnb = **(1-Epi**^{2})-1

Where n is the number of individuals ëtudiës. The expected unbiased hëtërozygosity is estimëed per locus and per population.

The different hërentes hëtërozygoties have ëtë dëterminatedë using Microsatellite Analyzer (MSA) software version 3.15 (Dieringer and Schlotterer, 2003).

> Inbreeding coefficient or measure of hëtërozygosity deficit (FIS)

This paramëtre represents the ratio of plus or minus heterozygosity observed in relation to the expected hëtërozygosity (HE) under the Hardy-Weinberg hypotheses. This new parameter defined by Wright (Wright, 1965) is called the fixation index (F) of individuals within subpopulations (s). In principle, it corresponds to the unbiasedë (f) estimator of Weir and Cockerham (1984) or dëficit in hëtërozygotes. It is also called the panmixy variance and is calculated as follows:

FIS = **HE-HOHE=** 1 - **HOHE**

It varies between -1 and +1. Equal values therefore correspond to an excess of hëtërozygotes, positive values to a deficit in hëtërozygotes and the zero value corresponds to the Hardy-Weinberg equilibrium. It is interesting to note that -1 can only be achieved by a population where all individuals are heterozygous for the same two al^les, whereas +1 only means that there are no hëtërozygotes and therefore all individuals are homozygous for the ëtudiës loci. Furthermore, it turns out that a good number of factors contribute to this discrepancy: inbreeding, derivation, selection, differentiation, etc. (De Meeus, 2012).

In order to test for local panmixy (FIS test), the alleles present in each sub-sample are randomly reassociated within these sub-populations and in all sub-populations. We then measure the overall FIS (average over all sub-samples and loci) (f estimate from Weir and Cockerham, 1984). This process is rëpëtë several times, which makes it possible to obtain the distribution of FIS generated under the hypothesis of local panmixy (HO) (De Meeus, 2012).

The FIS is thus estimated per population by considering all loci after 1000 permutations of the al^les within the populations; significance is considered when the percentages of rëpëtions present a value of the FIS infëer than the actual one, i.e. greater than 95%. The FIS values per locus for all populations were determined by the Jackknifing method on all loci and the confidence intervals by the Bootstrapping method on all loci. Significance of Fis values is inferred at the 95% confidence interval.

The Weir and Cockerham (1984) inbreeding coefficient (FIS) was calculated using FsTAT software version 2.9.3.2 (Goudet, 2002).

> Genetic balance

> Hardy-Weinberg equilibrium test (EHW)

Deviations from Hardy-Weinberg equilibrium were tested for each locus considering all populations and for each population, all loci combined. These exact tests were performed using the GENEPOP version 4.2.2 program (Rousset, 2008) by the Markov chain method with exact estimation of the Chi2 p-values by the Fisher method (fixed parameters are: dememorization = 10 000, batches = 20, 1 000 000 iterations per batch). The null hypothesis is that the populations are in Hardy-Weinberg equilibrium. When deviations from this equilibrium are observed, we agree that at least one of the hypotheses has not been verified.

> Genetic differentiation

The simplest parameters for evaluating genetic diversity between breeds using microsatellite data are the estimators of genetic differentiation or fixation indices. These are :

> FST (9) by Weir and Cockerham (1984)

FsT values or the population subdivision effect measure the reduction in heterozygosity in subpopulations related to differences in mean allelic frequencies. Also appelë the co-ascendency coefficient, 9 from Weir and Cockerham (1984) or "fixation index", the FST is dëfmi as ëbeing the correlation of gametes within subpopulations with respect to gametes drawn randomly from the whole population (all subpopulations included). It is calculated using the expected mean hëtërozygosity of the subpopulations and the expected hëtërozygosity of the total population. The FST is always positive and lies between 0 and 1. 0 corresponds to panmixy (mating occurs at random, no genetic divergence within the populations, hence no differences between the ancient frequencies of the subpopulations) and 1, complete isolation (all the subpopulations concerned are panmictic and totally isotropic). Thus, if the subpopulations share the same ancient frequencies (zero variance), this deficit is zero (no deviation from Hardy-Weinberg equilibrium); whereas in the case where the FST is positive, and all the more so as the ancient frequencies differ between subpopulations up to a maximum value of 1, when each subpopulation is fixed for one of the alleles present (maximum variance). This phënomëne gënërë by a variance between ancient frequencies is called the Wahlund effect (Wahlund, 1928), i.e. the dëficit in hëtërozygotes due to population structuring. The Wahlund effect, which results in a reduction in the heterozygosity of sub-populations relative to the total population, is observed when there is a divergence in ancient frequencies between sub-populations. This divergence is generally induced by the effect of gëtic dërive, which tends to cause ancient frequencies to diverge between different subpopulations (De Meeus, 2012).

FST values were thus measured by locus for all populations using the Fstat version 2.9.3.2 program (Goudet, 2002) by the Jackknifing method on all loci and the confidence intervals by Bootstraping on all loci. The significativitë of the values has ëtë dëduced according to a 95% confidence interval. However, we usedë Genetix software version 4.05.2 (Belkhir *et al.,* 2004) to dëerminate the FST values per population pair after 1000 permutations. Another measure of gënëtic diffërenciation analogous to the FST, the 9RH of Robertson and Hill (1984) was also estimëed by population pairs using the same procëdure. This measure takes into account the size of the alleles and not the ancient frequencies. Alleles of close size are more likely to have a close common ancestor, in the case where microsatellites follow the strict mutation model, the sMM. The significance of FST and 9RH values between pairs of populations has ëtë considërëe when the percentages of rëpëйоп3 presented a value of FST or 9RH lower than the real one, i.e. supërieure than 95%.

> FIT (F) by Weir and Cockerham (1984)

The FIT measures the homozygosity of individuals in the total population or the deficit in overall hëtërozygotes. When the subpopulations are not panmictic, in this case the overall heterozygote deficit may result from two effects: the Wahlund effect and the effect of non ateatory crosses in the subpopulations. Both FIT and FIS vary between -1 and 1 (De Meeus,

2012).

The ITF (F) was thus measured per locus for all populations using the FSTAT version 2.9.3.2 program (Goudet, 2002) by the Jackknifing and Bootstraping methods on all loci. The significance of the values was inferred according to a 95% confidence interval.

> General flow, Nm (Wright, 1969)

The gënëtic differentiation between populations is favoured by the gënëtic drift and limited by the gënëtic flows between populations. Knowledge of the STF can be used to infer the number of migrant individuals (the product Nm, with N: the size of each sub-population and m: the actual number of migrants) in a sub-population. Using the formula Nm = (1-FST)/4FST. Gene flow (Nm) between population pairs has ëtë dëterminë using Genetix software version 4.05.2 (Belkhir *et al.,* 2004). However, the estimate of gene flow per population pair ëtë dëduced from the FST according to the relationship dëcëcëdecribed previously.

> Mëthods for reconstructing phylogënëtic trees

To infërer parentë relationships between different populations following the evolutionary process, we used the Neighbour Joining (NJ) methods of Saitou and Nei (1987). In principle, in distance mëthods, the ëvolutive distance is calculated for all pairs of populations and the phylogenetic tree is constructed from the distance matrix. In addition, the reliability of the dendrogram nodes is measured using the bootstrap method (Felsenstein, 1985). The reconstruction of the phylogënëtic trees was carried out using the Neighbor-Joining and UPGMA methods implemented in the Populations version 1.2.28 program (Langella, 1999). The population tree was constructed from the distance matrix of shared alleles. The reliability of the nodes of the constructed trees was tested using the bootstrap method with 10000 rëë samples. Visualization and graphical reproduction of the dendrograms were performed using MEGA version X software (Kumar *et al.*, 2018).

> Genetic structure of populations

> Dëmographic bottleneck analyses

It has been shown that the notions of effective population size, bottleneck and conservation biology are closely linked. The effective population size, also known as the genetic population size and generally rated Ne, is thought to represent the rate at which a population loses its genetic diversity through genetic drift. A population undergoing a sharp reduction in population size (bottleneck) will tend to show a simultaneous reduction in the number of alleles per locus and in their genetic diversity (De Meeus, 2012). During a bottleneck, it is noted that the loss of alleles occurs faster than the decrease in genetic diversity. As a result, a population that has undergone a recent bottleneck will have a genetic diversity (He) greater than that expected at mutation/derivative equilibrium (Heq), given the number of observed alleles (k) under the assumption of a constant population size (Baker, 2005). In a population at mutation/derivative equilibrium whose size has not varied for a reasonable time, there is as much chance of observing an excess as a deficit of genetic diversity, compared with what is expected at the different loci. In order to detect whether the number of excesses observed significantly exceeds what is expected under this null hypothesis, three tests can be used, but the most convenient and effective is the Wilcoxon test (De Meeus, 2012).

Excess genetic diversity (He > Heq) has been demonstrated for loci evolving under the allelic infinity model (AIM). If the loci evolve under the strict stepwise mutation (SMM) model, there may be situations where the excess diversity is not observed. However, few loci follow the SMM model, and as soon as they become slightly mutated in favour of the IAM model, they show an excess of heterozygosity due to some genetic bottleneck in the past. Given that few loci follow the SMM model, the two-phase mutation model (TPM) is considered to be the most appropriate. The TPM is intermediate between the SMM and IAM models (Baker, 2005).

Several mutation models can be used depending on the situation. But according to De Meeus (2012), it is preferable to use all three mutation models simultaneously. This is because it has been shown that, if a bottleneck has actually occurred, it will be detected very strongly with the IAM hypothesis, moderately with the TPM and weakly with the SMM. However, if there is no bottleneck, but the population is structured into small sub-populations, a bottleneck signature may be falsely detected with IAM, but exceptionally (if ever) with TPM and never with SMM (De Meeus *et al.,* 2010). Two methods implemented in the Bottleneck software version 1.2.02 (Cornuet and Luikart, 1996) have been developed to determine whether populations are likely to have undergone recent bottleneck events.

> Wilcoxon test (Cornuet and Luikart, 1996)

The Wilcoxon test has been found to be more efficient and robust for testing excess heterozygosity relative to that expected at mutation-derivative equilibrium based on allelic frequencies (Mahmoudi *etal.,* 2012). To do this, we check the null hypothesis that all loci are in mutation-derivative equilibrium according to the different models generated. Therefore, if any deviation occurs, we obtain an excess of heterozygosity with a significant p-value. The default parameters are: 70% of SMM mutations in TPM and 30% of TPM involving the addition or removal of more than one microsatellite motif with a variance of 30 (De Meeus, 2012). Estimates were made on 1000 replications. The p-values for all populations were obtained using the generalized binomial test of Teriokhin *et al.* (2007) implemented in MultiTest version 1.2 (De Meeus *et al.,* 2009). For the test, we set k = 4 populations, giving a maximum probability of a = 0.0999.

> Mode-shift indicator test (Luikart *et al.,* 1998)

A second method, an L-shaped graphical representation of the "mode-shift" indicator developed by Luikart *et al* (1998), has also been used to test recent demographic bottlenecks. This method implements a qualitative descriptor of the distribution of allelic frequencies ("mode-shift" Indicator) to discriminate bottleneck populations from stable populations. In a population at mutation-derivative equilibrium (with an effective size, which has remained constant in the recent past), there is approximately an equal probability that a locus will show an excess of genetic diversity or a deficit of genetic diversity (Baker, 2005). Indeed, the loss of rare alleles in bottleneck populations is detected when one or more allelic classes share a high number of alleles than rare allelic classes (Luikart *et al.*, 1998). In the allelic frequency spectrum visualised by the qualitative graphical method, the microsatellite alleles are organised into 10 allelic classes. This makes it possible to check whether the distribution of allelic frequencies follows the L shape of the 'mode-shift' indicator, where low-frequency alleles (0.010.1) are the most abundant.

> Genetic assignment of individuals by the GeneClass programme

Gënëtic assignment tests are used to determine the probability of an individual belonging to its source population (origin). Cornuet *et al* (1999) suggestedërë a new population assignment approach (exclusion-simulation approach) which calculates for each individual, its probability of belonging with respect to each sampled population. In principle, the exclusion-simulation approach proposed by Cornuet *et al* (1999) calculates the probability that an individual belongs to a population by simulating 10,000 genotypes. For example, if an individual's

genotype is observed once out of 10,000 simulated genotypes, the probability that this individual belongs to a population is p = 0.001. This approach therefore makes it possible to exclude the populations of origin of the individuals. The p-value threshold is set according to a certain exclusion requirement (generally between 0.05 and 0.001). If the individual's probability is lower than the threshold set for that population, then the individual does not originate from that population. Therefore, an individual is considered to be correctly assigned to a population when it is excluded from all the others with a highly significant probability (p<0.001), except the one from which it originates.
Genetic assignment tests were carried out using the GeneClass version 2.0 program (Piry *et al.,* 2004) using the allelic frequency method of Paetkau *et al.* (1995). Individuals were assigned under the correlation of two hypotheses: the populations were assumed to be in Hardy-Weinberg equilibrium and the loci in linkage equilibrium (Cornuet *et al.,* 1999).

> Inference of the genetic structure of populations using a Bayesian approach with the Structure programme

In order to search for the occurrence of independent genetic groups (K) and to correctly assign each individual to the population where its genotype has the highest probability of occurring, we used Structure software version 2.3.4 (Pritchard *et al.,* 2000). This programme uses the Bayesian Markov Chain Monte Carlo (MCMC) approach, which is based on the clustering model, to infer the genetic structure of populations and check the correct assignment of individuals to their population of origin according to a certain probability Q (multidimensional vector representing the proportion of ancestry for all members of a population). According to Pritchard *et al* (2000), these individuals represent a тёlапде of K unobserved populations. The analysis under the Bayësian approach is performed under the assumption that: loci are in linkage equilibrium and populations are in Hardy-Weinberg equilibrium. This method has the advantage of providing reliable assignment results with a modest number of loci. The genetic structure of local populations was inferred by assuming the 'admixture' modële for ancestry; whereas for allelic frequencies, we used the 'correlated allelic frequencies' modële. This allows individuals to have several origins (mixed ancestry). According to Falush *et al* (2003), in the 'correlated allele frequencies' modële, allele frequencies from different populations are likely to be similar due to migration or shared ancestral polymorphism. Moreover, according to Falush *et al* (2003), this model, like the one using biallelic loci, assumes that all the populations diverged from a common ancestral population during the same period and that they have undergone different amounts of genetic drift (due to differences in the effective size of the populations) since their divergence. On the basis of these assumptions, and given a pre-assigned number of clusters (K), the programme uses the MCMC algorithm to calculate the natural logarithm *of the a posteriori* probability of clusters (K) in a population, given the observed genotype composition G (Ln Pr (K/G)). The latter is directly proportional to the natural logarithm of the probability (Pr) of the observed genotype composition (G) given the number of clusters (K) preassigned in the Structure programme dataset.
The analysis was carried out by assigning individuals to their reference population (sample population). The posterior probability values of K (log likelihood; lnL) are estimated with a burn-in period of 104 iterations, followed by 106 iterations of MCMC. For each value of K (1<K<8), 20 independent replications (runs) are performed to check the consistency of the estimates across iterations. The individual alpha admixture parameter was the same for all clusters and with uniform prior information. The K applied to the dataset will be the one for

which the posterior probability Ln P(D) is maximised. The proportion of ancestral genome (Q) was also deduced. The results generated by Structure were submitted to the Structure Harvester web site programme (Earl and VonHoldt, 2012) to determine the most probable number of gene groups (K) using the L(K) likelihood method and Evanno's method, by determining the modal distribution of DeltaK (AK) values (Evanno *et al.,* 2005).

II.3 Molecular characterisation of West African cattle

II.3.1. Sampling

In addition to the bull samples from Burkina Faso, 445 samples belonging to nine (9) cattle breeds and three (03) mëtis populations representing four (04) main cattle subtypes found in West Africa were sampled in this study. These ë samples were collected in four different countries (Figure 17). Eight (8) cattle of the N'dama breed, representing long-horned bulls, were sampled in Mali. In addition to the Burkina taurins, specimens of short-horned taurins were collected in Benin (Borgou: 14 and Lagune: 23). Bulls with massive horns were collected in Niger (Kuri: 36). The Zebu type was represented by five breeds sampled respectively in Benin (Zebu Peul: 34), Burkina Faso (Bororo: 20 and Goudali: 37) and Niger (Arabe: 52 and Bororo: 109). Sanga (mixed-breed) cattle were represented by Borgou x Zebu Peul crosses from Benin (23), Kouri x Arab (29) and Kouri x Bororo (40) from Niger. The main criteria that prevailed for this sampling were the unrelatedness of the subjects, their purity (for the identified breeds) and the random nature of the choice, with one bovine being sampled per herd. Details of the sampling are given in Table IV.

Table III: Description of cattle samples West Africa

	Denomination	Type	Country	Workforce	Provinces	Villages
1	Zebu Peul	Zebu	Benin	34	3	14
2	Bororo Burkina	Zebu	Burkina Faso	20	1	5
3	Boudali	Zebu	Burkina Faso	37	1	4
4	Pure Arabic	Zebu	Niger	52	1	12
5	Bororo Diffa	Zebu	Niger	29	2	3
6	Bororo site Kouri	Zebu	Niger	50	2	3
7	Bororo Maaoua	Zebu	Niger	30	3	5
8	Borgou x Zebu	Sanga	Benin	23	2	4
9	Kouri x Arabic	Sanga	Niger	29	1	3
10	Kouri x Bororo	Sanga	Niger	40	1	3
11	Borgou	Taurin	Benin	14	4	4
12	Kouri Pur	Taurin	Niger	36	1	12
13	Lagoon	Taurin	Benin	23	7	8
14	Gourounsi Nahouri	Taurin	Burkina Faso	40	1	11
15	Gourounsi Sanguie	Taurin	Burkina Faso	39	1	7
16	Lobi	Taurin	Burkina Faso	41	1	9
17	N'Dama	Taurin	Mali	8	2	3

Blood samples were taken by jugular vein puncture and collected in EDTA tubes. The collected samples were kept at +4°c pending extraction of 1 gënomic DNA. DNA was ële extracted from whole blood using the MasterPure DNA purification kit (Biozym, Illumina Inc., USA). The ёсЬатШоп3 DNA then ële storedës at 4 ° C until PCR amplification. Twenty-seven (27)27 microsatellite markers selected from those recommendedës by FAO

(2011) were ëlës used. The primers were conjugated with one of four fluorescent dyes (FAM, HEX and ATTO550). The PCR reaction was performed under the following conditions: initial dënaturation for 5 min at 95°C, followed by 35 cycles of dënaturation at 95°C for 30 seconds, 1 minute of hybridization at spëcific temperatures for each primer, extension at 72°C for 1 minute and final extension at 72°C for 10 minutes. PCR products were then ëlësubjected to ëlectrophorëse after multiplexing in an ABI3100 automatic DNA sequencer (Applied Biosystems, USA) using ROX500 (Applied Biosystems, USA) as an internal marker. All 27 microsatellite loci were ëlë multiplexedës into six panels for gënotyping (Table II). The gënotypes were then ëlë extracted using GENEMAPPER software.

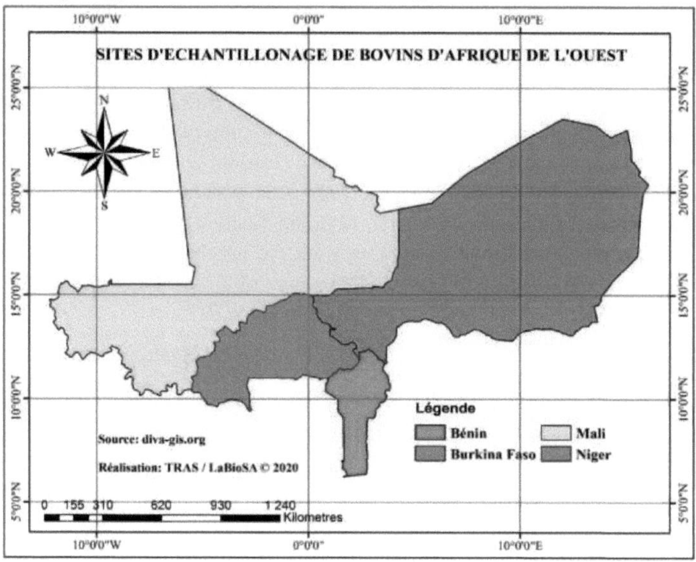

Figure 17: West African cattle sampling sites (Tapsoba, 2020)

II.3.2. Statistical analysis

The presence of null alleles was checked using Microchecker software version 2.2.3 (Oosterhout et al., 2004). Basic diversity indices such as the number of observed alleles, observed and expected heterozygosity, F_{IS} heterozygote deficit and Wright's F-statistics were calculated using MICROSATELLITE ANALYZER (MSA) version 3.15 (Dieringer and Schlotterer, 2003). Deviations of heterozygosities from Hardy-Weinberg equilibrium (HWE) were estimated by: (1) calculating the degree of intrapopulation reduction in heterozygosity due to inbreeding (F_{IS}) according to Wright (1951); and (2) exact tests of the excess and deficit of heterozygotes for each marker and in each breed, as implemented in GENEPOP (Raymond and Rousset, 1995). The allelic distances between populations and between individuals were also calculated using the MSA software (Dieringer and Schlotterer, 2003). The pairs of allelic distances between individuals were used to construct the phylogenetic tree following the Neighbor Joining method using PHYLIP version 3.5 (Felsenstein, 1993). The tree was visualised using MEGA version 6.0 (Tamura *et al.*, 2013). To verify the recent history of cattle breeds in Niger, assignment probabilities were calculated using likelihood (Paetkau *et al.*, 1995) and Bayesian methods (Rannala and Mountain, 1997, Baudouin and Lebrun, 2001) using GeneClass 2 software (Piry *et al.*, 2004). The scatter plot of likelihood estimates (Baudoin and Lebrun, 2001) for individuals of each cattle breed was constructed using SPSS version 26.0 (IBM corp, 2019). The Nigerian cattle populations were tested for the equilibrium of mutation derivatives using three statistical approaches (sign test, standardized differences test and Wilcoxon sign rank test) according to the 3 microsatellite marker evolution models using the BOTTLENECK programme (Piry *et al.*, 1999).

CHAPTER III

RESULTS

III.1. Morphobiometric characterisation

The interactions between the different measurements were analysëes through the two-dimensional scatter plot (Figure 19 and 20). The first two principal components of the 2014 PCA graph contributed 66.6% of the total variability, while those of 2018 grouped 69% of the total inertia (Table V). Strong similarities were observed between these two graphs. From 2014 to 2018, most of the variability was carried by eleven (11) variables (head length, height at sacrum and hips, head width, horn length, chest depth, thoracic përimëtre, skull width, skull length, sacpulo-ischial length and chest width). These variables were positively and strongly correlated (>0.6) to the first dimension, which capitalised 49.2 and 61.6% of the variability in 2014 and 2018 respectively (Table V). However, between 2014 and 2018, a different trend was observed for certain variables. The length of the face, pelvis and ears, the circumference of the muzzle, the width of the hip and the width at the ischium were moderately to negatively correlated with the first dimension in 2014, whereas in 2018 they were strongly correlated with the same dimension. Tail length was negatively correlated with the first dimension in both 2014 and 2018. The second dimension was positively correlated with pin length and hip width in 2014 and 2018 respectively. In addition, the second dimension of the 2014 PCA graph had more negative correlations and higher absolute values compared to the second dimension of the 2018 graph.

Table IV: Correlation of variables in the construction of the first two principal components

2014			2018	
Parameters	**CP 1**	**CP 2**	**CP1**	**CP2**
Proportion of variations	Cp 1	CP 2	CP1	CP2
Head length	49,2	17,4	61,6	7,4
Height to sacrum	0,93	-	0,96	-0,12
Height at withers	0,91	-	0,93	-
Head width	0,9	0,09	0,89	-
Length of horns	0,86	-0,35	0,9	-0,19
Depth of chest	0,82	-	0,83	-0,12
Thoracic perimeter	0,81	0,2	0,79	0,26
Width of skull	0,81	-	0,81	0,28
Skull length	0,81	-0,34	0,87	-0,13
Scapulo-ischial length	0,8	-0,46	0,88	-
Chest width	0,78	0,21	0,82	-
Face length	0,61	-	0,73	-0,16
Length of pool	0,58	0,59	0,86	-
Ear length	0,57	0,67	0,8	-
Muzzle circumference	0,54	-0,62	0,86	-0,16
Hip width	0,35	0,14	0,62	0,28
Width at ischium	0,31	0,34	0,47	0,71
Tail length	0,12	0,84	0,46	0,46

-CP: Main Component

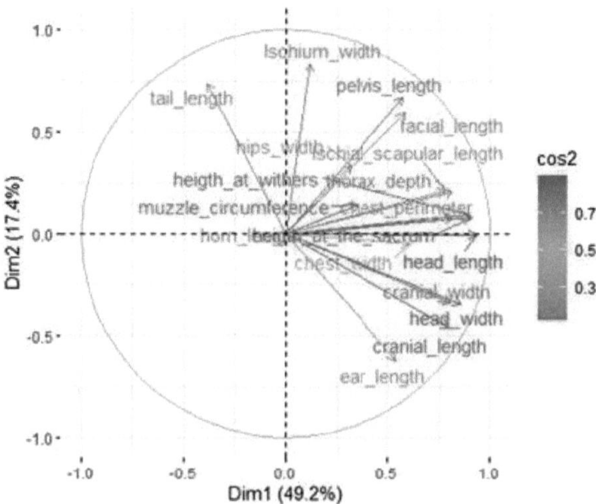

Figure 18: Correlations between the different variables and the different principal components 1 and 2 (year 2014).

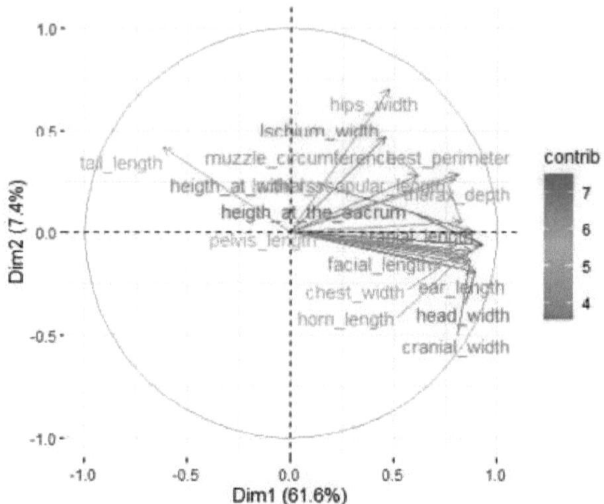

Figure 19: Correlations between the different variables and the different principal components 1 and 2 (year 2018)

The graph showing the distribution of individual variability constructed using the 2014 data shows 4 representative clusters for each of the four (4) breeds studied (figure 20). The Gourounsi de Nahouri (GourN) and the lobi form one cluster, while the clusters formed by the Gourounsis du Sanguie taurins and the Peul zebu are quite distinct. With reference to individual characteristics, the first dimension of the PCA graph of individuals allows a separation of individuals into two groups, namely those characterised by high morphological

values on the one hand, and those characterised by low body measurements on the other.
The first group is mainly made up of Peul zebus characterised by high values for skull length, head width, skull width, head length, height at sacrum, height at withers, horn length, ear length, thoracic perimeter and scapulo-ischial length. In addition, tail length and width at ischium are not as important in this breed.
The second group consists of animals that share low values of variables such as scapulo-ischial length, pelvic length, face length, height at sacrum, height at withers, thorax depth, head length, horn length, thoracic perimeter and ischium width. On the graph, this group is represented by the overlap between the Lobi taurin clusters and that of the Gourounsi populations of the Nahouri.
The third group consisted mainly of Gourounsi Nahouri. Individuals in group 3 showed high values for tail length, width at ischium, pelvic length, face length, and hips; low values for ear length, skull length, skull width, head width, chest width, height at sacrum, horn length, height at withers, and thoracic përimëtre.

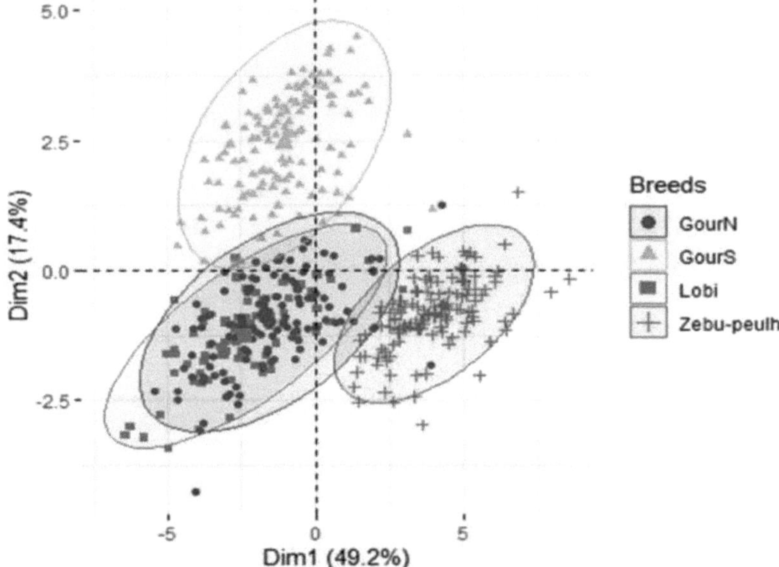

Figure 20: Projection of individuals of different races (Lobi, Zebu peul, and Grourousni Nahouri and Sanguie) in the two-dimensional plane (1-axis2) of the year 2014

Compared to the 2014 graph of individuals, the 2018 graph identified four (4) different clusters belonging to each of the ëtudiëes breeds (Figure 21). In 2018, the three bull breeds formed a single cluster, while the Fulani zebu formed a distinct cluster that was well individualised from the first. The first dimension contrasts full-bodied individuals with slim individuals. Individuals in the first cluster have high values for head length, chest width, face length, height at sacrum, skull length, skull width, height at withers, horn length, pelvic length and scapulo-ischial length and low values for tail length. Group 2 is characterised by high values for thoracic perimeter, depth of chest, circumference of muzzle, width of head, height at sacrum, length of head, height at withers, width of hip, length of ears and length of face.

However, group 2 is characterised by a short tail. Group 3 has high values for tail length and low values for head length, height at sacrum, face length, height at withers, head width, skull width, skull length, scapulo-ischial length, pelvic length and horn length.

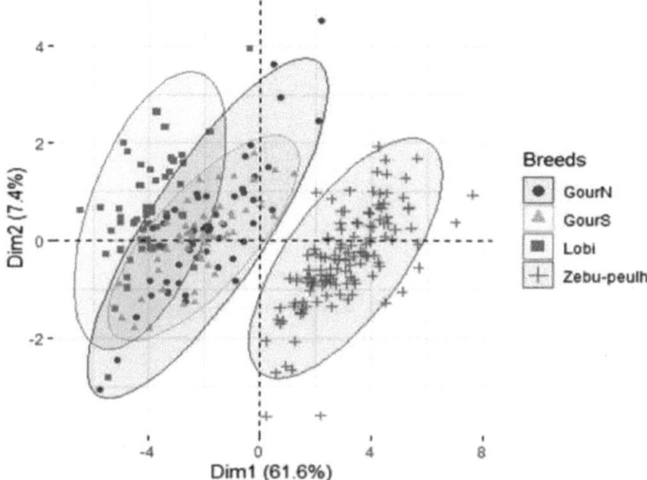

Figure 21: Two-dimensional graph showing the dispersion of individual variability (2018)

For a precise description of the variability in the two datasets, a bottom-up hierarchical classification was performed. In addition, this classification was optimised by determining the best possible number of groups (k) using the method proposed by Kassambara *et al.* (2008). In 2014, the optimal number of groups was 3 (Figure 22), while in 2018 there were 4.

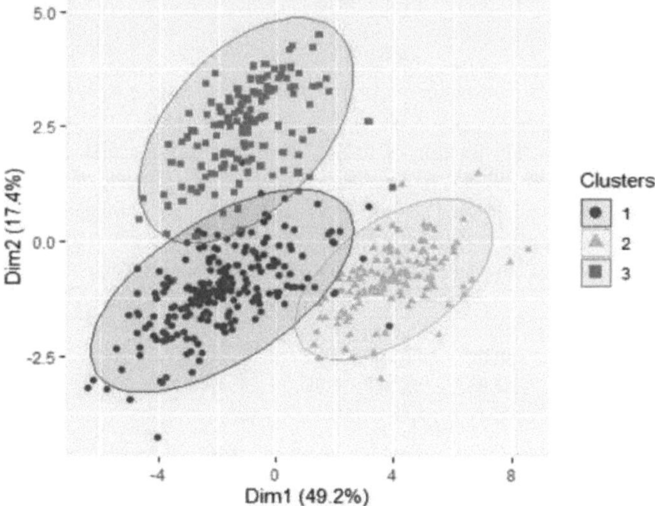

Figure 22: Bottom-up hierarchical classification chart (2014)

In 2014, the first cluster consisted mainly of Lobi (98.15%) and GourN (94.80%), with a small proportion of GourS (7.25%). The second cluster was made up almost exclusively of GourS (91.30%). All Fulani zebu were found in cluster 3 (89.79%). However, a few individuals of the Gourounsi (GourS) (6.08%) and Lobi (2.04%) breeds were also found in this last cluster (table VI).

Table V: Composition of clusters resulting from the classification (2014)

	Cluster 1			Cluster 2			Cluster 3
Breeds	Cla/Mod	Mod/Cla	Cla/Mod	Mod/Cla	Cla/Mod	Mod/Cla	
GourN	89.1558		.083	.10	3.057 .75		6.80
Lobi	94.8036	.871	.30		0.763 .90		2.04
GourS	7.	255.0591	.30		96.18		1.451 .36
Zebu peul		000			0		10089 .79

-Cla/Mod: Class/Mode; -Mod/Cla: Class/Mode

The 2018 hiërarchical classification of animals shows 1 existence of 4 groups (Figure 23).

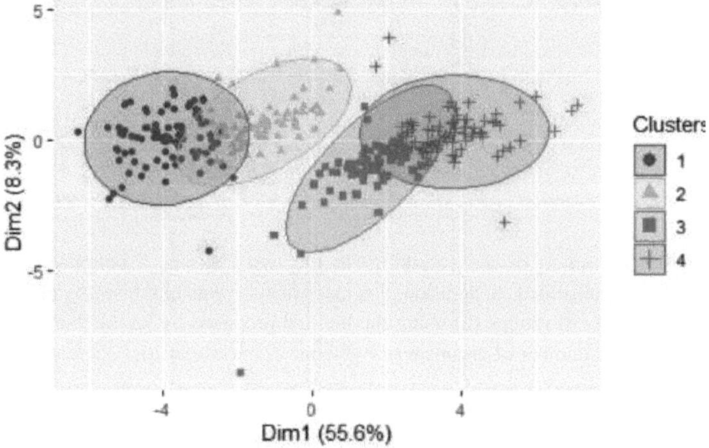

Figure 23: bottom-up hierarchical classification graph (2018)

With the exception of a tiny fraction of Fulani zëbus (0.65%), the first cluster is made up exclusively of Lobi taurin (92.45%). The second cluster is Oquitableë composed of GourN and GourS taurin (60 vs 62.96%). The third and fourth groups are dominated by Fulani zëbus with a few Gourounsi taurins (GourN and GourS) (Table VII).

Table VI: Composition of clusters resulting from the classification (2018)

	Cluster1		Cluster 2		Cluster 3		Cluster 4	
Breeds	Cla/mod	Mod/cla	Cla/mod	Mod/cla	Cla/mod	Mod/cla	Cla/mod	Mod/cla
GourN	0	0	60,00	44,11	0	0	4.00	2.,40
GourS	0	0	62,96	50,00	1,85	1,40	0	0
Lobi	92,45	56,32	7,54	5,88	0	0	0	0
Zebu Peulh	0,65	1,14	0	0	46,05	98,59	53,28	97,59

Linear discriminate analysis (LDA) was used to classify the animals according to their morphological similarities in their different breeds (Table VIII). The GourN accounted for 2.31% of misclassified individuals, including 1.54% classified as GourS and 0.77% as lobi.

98.7% of the lobi bulls were well classified. This population seems to share
some morphological similarities with GourN or 1.3% of misclassified lobi were affected. However, as in the PCA plot, no morphological similarities were evidenced by 1 ALD between lobi and GourS; 100% of GourS individuals had good ëtë classës.

Table VII: Classification (%) based on linear discriminate analysis (2014)

Predicted classes	GourN	GourS	Lobi	Zebu peul
GourN	97,69	1,54	0,77	0
GourS	0	100	0	0
Lobi	1,3	0	98,7	0
Zebu peul	0,75	0	0	99,25

In 2018, the proportion of individuals with morphological similarities between GourS and GourN was around 20% in each group. As in the 2014 ALD, none of the GourS individuals were misclassified as lobi. However, 7.27% of lobi individuals were ëtës misclassified as GourS and 5.45% as GourN (Table IX).

Table VIII: Classification based on linear discriminant analysis (2018)

Predicted classes	GourN	GourS	Lobi	Zebu peul
GourN	75,56	20	4,44	0
GourS	21,15	78,85	0	0
Lobi	7,27	5,45	87,27	0
Zebu peul	0	0	0	100

The least squares mean of the various measurements by breed and year is given in Table X. The effect of the breed-time interaction allowed evaluation of the morphological evolution of the different breeds studied over time. No significant variation between the measurements of the different years was found in the GourN individuals. However, with the exception of head length, thoracic përimëtre, chest width and muzzle circumference, all body measurements of GourS individuals significantly increasedë between 2014 and 2018. Similarly, in the Lobi breed, variables such as face length, horn length, height at withers, chest depth, scapulo ischial length, pelvic length, 60
Chest width, hip width and ear length increased significantly between 2014 and 2018.
Depending on their height at the withers, Burkina bulls can be divided into three groups:
small bulls (lobi), medium bulls (GourN) and large bulls (GourS).

Table IX: Average least squares and standard errors for body measurements (2018)

	GourN		GourS		Lobi	
parameters	[2014]	[2018]	[2014]	[2018]	[2014]	[2018]
Head length	$43,41 \pm 0,17^{a}$	$42,39 \pm 0,23^{a}$	$43,71 \pm 0,17^{a}$	$42,69 \pm 0,23^{a}$	$40,33 \pm 0,22^{a}$	$39,31 \pm 0,24^{a}$
Head width	$19,38 \pm 0,14^{a}$	$19,46 \pm 0,18^{a}$	$17,79 \pm 0,14^{a}$	$17,87 \pm 0,18^{b}$	$17,8 \pm 0,17^{a}$	$17,88 \pm 0,1^{a}$
Length Crane	$21,07 \pm 0,12^{a}$	$22,03 \pm 0,17^{a}$	$19,55 \pm 0,13^{a}$	$20,51 \pm 0,16^{b}$	$19,96 \pm 0,16^{a}$	$20,91 \pm 0,17^{a}$
Width Crane	$16,38 \pm 0,13^{a}$	$16,49 \pm 0,17^{a}$	$15,14 \pm 0,13^{a}$	$15,24 \pm 0,17^{b}$	$15,2 \pm 0,16^{a}$	$15,31 \pm 0,18^{a}$
Length Face	$22,34 \pm 0,15^{a}$	$20,03 \pm 0,2^{a}$	$24,83 \pm 0,15^{a}$	$22,52 \pm 0,2^{b}$	$20,5 \pm 0,19^{a}$	$18,19 \pm 0,21^{b}$
Length Horns	$17,49 \pm 0,5^{a}$	$18,55 \pm 0,66^{a}$	$21,37 \pm 0,5^{a}$	$22,43 \pm 0,65^{b}$	$14,06 \pm 0,62^{a}$	$15,11 \pm 0,68^{b}$
Withers	$95,69 \pm 0,43^{a}$	$100,94 \pm 0,58^{a}$	$97,46 \pm 0,44^{a}$	$102,72 \pm 0,57^{b}$	$88,18 \pm 0,55^{a}$	$93,43 \pm 0,6^{a}$

height						
Chest depth	49,75±0,24^{a}	50,25±0,33^{a}	50,96±0,24^{a}	51,46±0,32^{b}	49,33±0,31^{a}	49,83±0,34^{b}
Sacrum height	99,2±0,45^{a}	104,5±0,59^{a}	101,97±0,45^{a}	107,27±0,59^{b}	91±0,56^{a}	96,3±0,61^{a}
Scapulo-ischial length	103,6±0,69^{a}	111,01±0,93^{a}	115,62±0,7^{a}	123,03±0,91^{b}	109,45±0,87^{a}	116,86±0,96^{b}
Basin length	34,51±0,29^{a}	31,83±0,38^{a}	40,26±0,29^{a}	37,58±0,38^{b}	33,15±0,36^{a}	30,47±0,39^{b}
Width Ischions	11,92±0,16^{a}	10,54±0,21^{a}	15,15±0,16^{a}	13,77±0,21^{b}	12,41±0,2^{a}	11,04±0,22^{a}
Length Tail	98,63±0,86^{a}	95,9±1,15^{a}	108,87±0,86^{a}	106,14±1,14^{b}	88,38±1,08^{a}	85,65±1,19^{a}

	GourN		**GourS**		**Lobi**	
Perimeter Thoracic	134,7±0,72^{a}	136,56±1,22^{a}	134,85±0,72^{a}	134,56±1,17^{a}	131,06±0,97^{a}	132,18±1,18^{a}
Width Breast	15,02±0,18^{a}	14,84±0,3^{a}	15,66±0,18^{a}	15,06±0,29^{a}	16,59±0,24^{a}	12,02±0,29^{b}
Hip width	28,29±0,23^{a}	26,97±0,31^{a}	29,86±0,23^{a}	28,54±0,3^{b}	30,96±0,29^{a}	29,64±0,32^{b}
Circonference Snout	37,29±0,2^{a}	35,46±0,27^{a}	36,82±0,2^{a}	34,99±0,27^{a}	36,22±0,26^{a}	34,4±0,28^{a}
Ear length	37,29±0,2^{a}	35,46±0,27^{a}	36,82±0,2^{a}	34,99±0,27^{b}	36,22±0,26^{a}	34,4±0,28^{b}

^{a,b,c} averages with the same letters in the same column are not significantly different at $p < 0.05$

A multivariate analysis of variance revealed an increase in all morphological traits except head width, skull width, horn length, thorax depth and thoracic përimëtre (table XI).

Table X: Results of the multivariate analysis between 2014 and 2018 on bulls

Features	**Model**	**Dl**	**SDC**	**Average**	**Value of F**	**F value**
Head length	Year	1	219,3	219,25	30,799	4,56E-08
Head width	Residus	522	3716,1	7,11		
Crane length	Year	1	0,21	0,2	0,0536	0,817
Crane width	Residus	522	2014,24	3,85		
Length Face	Year	1	86,9	86,89	27,536	2,25E-07
Length of horns	Residus	522	1647,3	3,15		
Withers height	Year	1	0,04	0,035	0,0112	0,9158
Chest depth	Residus	522	1664,45	3,188		
Sacrum height	Year	1	774	773,96	112,78	<2,2e-16
Scapulo-ischile length	Residus	522	3582,3	6,86		
Basin length	Year	1	23,4	23,37	0,4572	0,4992
Ischium width	Residus	522	26689,8	51,13		
Length of tail	Year	1	234,8	234,76	35,886	3,90E-09
Thoracic perimeter	Residus	522	3415	6,54		
Features	**Model**	**Dl**	**SDC**	**Average**	**Value of F**	**F value**
Chest width	Year	1	1065,1	1065,09	44,81	5,62E-11
Hip width	Residus	522	12407,6	23,77		
Circumference Snout	Year	1	1989,1	1989,11	42,794	1,45E-10
Ear length	Residus	522	24262,9	46,48		

-Dl: Degree of freedom, -SDC: Sum of Squares

III.2 Molecular characterisation of taurins from Burkina Faso

A total of 3861 genotypes were generated from the 27 microsatellites used. All microsatellite loci were rĕvēlĕs polymorphic for 1 set of ĕcЬатШопз. The number of allbles observed per locus (Na) shows a large alial variability between loci. Indeed, 99.97% of the 81 microsatellite combinations presented more than 4 alleles per locus. A total of 219 alleles were observed. The Number of alleles per locus ët ranged from 3 (INRA035) to 14 (TGLA53) for an average per population of 6.04; 7.41 and 6.04 for the Gourounsi (Nahouri and Sanguie) and Lobi taurine populations respectively. The average PIC per population was 0.54, 0.62, and 0.66 for the Lobi, Gourounsi du Nahouri and Gourounsi du Sanguie taurine populations respectively. Most loci had PIC values greater than 0.5 except for 4 loci in the Gourounsi du Nahouri taurine population (SPS115, INRA035, ILSTS05, ETH185, INRA63), 2 in the Gourounsi Sanguie taurine population (SPS115, INRA035) and 9 loci (INRA035, SPS115, ILSTS05 ; TGLA122, ILSTS006, HEL13, ETH152, INRA023, ETH10) in the Lobi 5 bull population (Table XII).

Tableau XI: Allelic diversity and details of the microsatellites used

SizeGourounsiGourounsi

allelesNahouriSanguieLobi

Locus panel	Hybridization temperature	Dyes	pb	Na	PIC	Na	PIC	Na	PIC
CSRM60	60°C	FAM	89-107	8	0,70	9	0,71	7	0,61
CSSM66	60°C	FAM	177-197	8	0,75	10	0,62	6	0,55
HEL1	56°C	HEX	98-114	6	0,64	7	0,71	7	0,54
INRA63	56°C	HEX	173-183	5	0,56	5	0,52	4	0,50
BM1824	61°C	ATTO550	183-197	4	0,57	4	0,51	4	0,58
ETH152	60°C	FAM	185-199	6	0,69	6	0,67	4	0,45
			Allele size	Gourounsi Nahouri		Gourounsi Sanguie		Lobi	
HAUT27	54°C	HEX	140-150	6	0,64	6	0,60	5	0,58
INRA05	54°C	FAM	134-140	4	0,59	4	0,52	4	0,51
BM1818	60°C	HEX	256-272	7	0,78	9	0,83	7	0,78
ETH3	63°C	FAM	99-125	5	0,57	8	0,70	8	0,52
HEL9	56°C	ATTO550	155-175	9	0,77	9	0,82	8	0,65
ILSTS006	54°C	FAM	284-300	5	0,61	6	0,59	4	0,41
TGLA53	55°C	HEX	151-183	13	0,87	14	0,79	9	0,75
HAUT24	53°C	HEX	103-125	7	0,69	8	0,77	8	0,64
HEL5	54°C	FAM	148-164	8	0,74	7	0,67	7	0,70
INRA032	56°C	ATTO550	164-208	7	0,68	9	0,75	8	0,62
SPS115	61°C	FAM	243-255	4	0,18	5	0,31	4	0,20
ETH185	65°C	ATTO550	224-242	6	0,44	8	0,56	6	0,57
HEL13	54°C	HEX	182-194	5	0,58	5	0,59	4	0,43
ILSTS05	56°C	FAM	178-190	3	0,42	6	0,56	4	0,40
INRA035	60°C	FAM	99-117	4	0,19	5	0,44	3	0,18
TGLA126	54°C	HEX	114-128	7	0,64	7	0,76	7	0,72
BM2113	63°C	FAM	118-142	7	0,75	9	0,77	7	0,64
ETH10	61°C	FAM	207-223	6	0,66	7	0,73	7	0,47
ETH225	63°C	ATTO550	140-160	8	0,72	8	0,76	9	0,65
INRA023	58°C	ATTO550	199-219	6	0,63	8	0,73	7	0,46
TGLA122	58°C	HEX	134-174	9	0,65	11	0,74	5	0,40
Average	-	-	-	6,41	0,62	7,41	0,66	6.04	0,54

Na-number of alleles observed; PIC-Polymorphism Information Content, pb-Base pairs

The mean observed and expected hëtërozygosity ët values were 0.639 and 0.659, respectively, for all populations. Overall observed heterozygosity at each locus ranged from 0.05 (INRA035) to 0.92 (CSRM60), while overall expected heterozygosity varied from 0.2 (SPS115) to 0.9 (TGLA53). By population, the observed mean was 0.64, 0.7 and 0.58 for Gourounsi Nahouri, Gourounsi Sanguie and Lobi cattle respectively; while the expected overall heterozygosity was 0.67 and 0.71 for Gourounsi cattle (Nahouri and Sanguie) and 0.6 for Lobi. The overall mean FIS was 0.028 and ranged from -0.36 (CSRM60) to 0.73 (INRA035). The highest estimate of the inbreeding coefficient (FIS) within ëtudiëe populations was foundëe in the Gourounsi taurin population of Nahouri (0.06) and the lowest mean reported in the Gourounsi sanguie population (0.00). The Hardy-Weinberg equilibrium (HWE) test revealed 11 locus x race combinations, i.e. 13.58% of the total number of loci with significant deviations (P <0.05) from panmictic equilibrium (Table XIII). Among the populations, the greatest number of loci deviating from HWE equilibrium (P> 0.05) for heterozygosity deficit was observed in Gourounsi Nahouri cattle with 4 loci followed by Gourounsi sanguie and Lobi cattle with 2 loci each. In addition, the HWE test for heterozygosity excës revealed significant deviations in 2, 0 and 1 loci respectively in Gourounsi Nahouri, Gourounsi Tenado and Lobi cattle.

Tableau XII: Estimation of the heterozygote deficit and expected and observed heterozygosities

LOCUS	Gourounsi Nahouri			Gourounsi Sanguie			Lobi		
	Ho	He	FIS	Ho	He	FIS	Ho	He	FIS
CSRM60	0,84	0,75	-0,12	0,71	0,74	**0,04**	0,92	0,68	-0,36
CSSM66	0,71	0,79	0,09	0,63	0,67	0,05	0,67	0,63	-0,06
HEL1	0,76	0,69	-0,1	0,83	0,76	-0,09	0,51	0,59	0,13
INRA63	0,61	0,64	0,04	0,68	0,61	-0,11	0,5	0,58	0,13
BM1824	0,44	0,63	0,29	0,68	0,57	-0,2	0,69	0,67	-0,04
ETH152	0,61	0,75	**0,18**	0,55	0,73	0,25	0,56	0,55	**-0,03**
HAUT27	0,82	0,7	-0,18	0,7	0,66	-0,07	0,64	0,64	-0,01
INRA05	0,59	0,66	0,09	0,64	0,58	-0,12	0,56	0,58	0,03
BM1818	0,78	0,82	0,04	0,9	0,87	-0,05	0,78	0,82	0,04
ETH3	0,62	0,65	0,03	0,75	0,75	0	0,64	0,6	-0,08
HEL9	0,7	0,81	0,13	0,8	0,86	0,06	0,74	0,69	**-0,08**
ILSTS006	0,83	0,68	**-0,23**	0,68	0,65	-0,04	0,38	0,5	0,22
TGLA53	0,76	0,9	**0,15**	0,74	0,83	**0,1**	0,62	0,79	0,22
HAUT24	0,82	0,74	-0,11	0,74	0,81	0,08	0,76	0,7	-0,1
HEL5	0,73	0,79	0,07	0,74	0,72	-0,03	0,76	0,75	-0,01
INRA032	0,68	0,73	0,07	0,86	0,79	-0,09	0,62	0,68	**0,09**
SPS115	0,21	0,2	-0,08	0,38	0,33	-0,15	0,21	0,22	0,01
ETH185	0,51	0,49	-0,07	0,59	0,6	0,02	0,55	0,62	0,1
HEL13	0,53	0,65	**0,19**	0,59	0,65	0,09	0,49	0,53	0,07
LOCUSGourounsi		**NahouriGourounsi**			**SanguieLobi**				
ILSTS05	0,54	0,54	0	0,71	0,64	-0,11	0,39	0,49	0,19
INRA035	0,05	0,2	**0,73**	0,36	0,51	0,29	0,18	0,2	0,14
TGLA126	0,69	0,7	0,01	0,76	0,8	0,05	0,79	0,77	-0,04
BM2113	0,65	0,79	0,18	0,78	0,81	**0,02**	0,64	0,7	0,08
ETH10	0,57	0,71	0,19	0,79	0,78	-0,02	0,53	0,51	-0,04
ETH225	0,89	0,78	**-0,16**	0,74	0,8	0,08	0,59	0,71	0,17

INRA023	0,61	0,68	0,09	0,76	0,77	0	0,54	0,52	-0,04
TGLA122	0,64	0,71	0,1	0,84	0,77	-0,1	0,44	0,43	-0,02
Average	0,64	0,67	0,06	0,7	0,71	0	0,58	0,6	0,02

Ho- hĕtĕrozygosity observed; He hĕtĕrozygosity expected; Fis-Estimation of the deficit in hĕtĕrozygotes

Overall F-statistics for the populations studied are presented in Table XIV. Fixation index (F_{ST}) values per locus ranged from 0.00 (HAUT27) to 0.9 (ILSTS05) with a mean of 0.04 for all loci. Mean F_{IT} estimates were 0.06 and ranged from -0.09 (CSRM60) to 0.41 (ILSTS05).

Tableau XIII: F-statistics

Locus	**FST**		**FIT**	**FIS**
CSRM60	0,04		-0,09	-0,13
CSSM66	0,04		0,08	0,04
HEL1	0,03		0,01	-0,02
INRA63	0,03		0,04	0,02
BM1824	0,08		0,22	0,15
ETH152	0,01		-0,06	-0,08
HAUT27	0		0,01	0,01
INRA05	0,02		0,03	0,01
BM1818	0,02		0,02	-0,01
ETH3	0,05		0,09	0,05
HEL9	0,03	0		-0,03
ILSTS006	0,04	0,2		0,16
TGLA53	0,06		0,04	-0,03
HAUT24	0,04		0,05	0,01
HEL5	0,01		0,03	0,02
INRA032	0		-0,07	-0,08
SPS115	0,01		0,04	0,03
ETH185	0,01		0,13	0,13
HEL13	0,05		0,07	0,02
Locus	**FST**		**FIT**	**FIS**
ILSTS05	0,09		0,41	0,36
INRA035	0,02		0,04	0,02
TGLA126	0,02		0,11	0,1
BM2113	0,05		0,1	0,05
ETH10	0,04		0,08	0,04
ETH225	0,06		0,08	0,02
INRA023	0,08		0,08	0
Total	0,04		0,06	0,03

Carrying the comparison between pairs of populations, maximum differentiation was observed in the Lobi-Gourounsi Sanguie pair (0.05) while the Gourounsi Sanguie-Gourounsi Nahouri pair was the least differentiated (0.02) (Table XV). In addition, the flux of gĕnes (Nm) per population pair based on F_{ST} values was calculated for all possible population pairs. As with F_{ST}, gene flow was high between Gourounsi populations (Gourounsi Sanguie and Gourounsi Nahouri) and lower for the Lobi-Gourounsi Tenado pair (Table XV).

Table XIV: FST values between population pairs (lower triangle) and gene flow (Nm) (upper

triangle)

Populations	FGN	FLB	FGT
FGN	-	8,08	12,25
FLB	0,03	-	4,75
FGT	0,02	0,05	-

-FGN: Gourounsi Nahouri, FGT: Gourounsi Sanguie; FLB: Taurins Lobi

In order to inférer the cryptic gënëtic structure of bullfighting populations in Burkina Faso, individuals from the different populations studied were clustered using a Bayesian approach. From this analysis, it emerged that K = 2 best describes the bull breeds of Burkina Faso (Figure 24). Thus, assuming that the most likely number of ancestral populations is two (02) (Figure 25), all the bull breeds in Burkina Faso are inbred with zebu genes. However, the Lobi breed appears to have a lower level of introgression than the Gourounsi cattle populations, whose level of introgression varies between 20 and 40% depending on the individual (Figure 24).

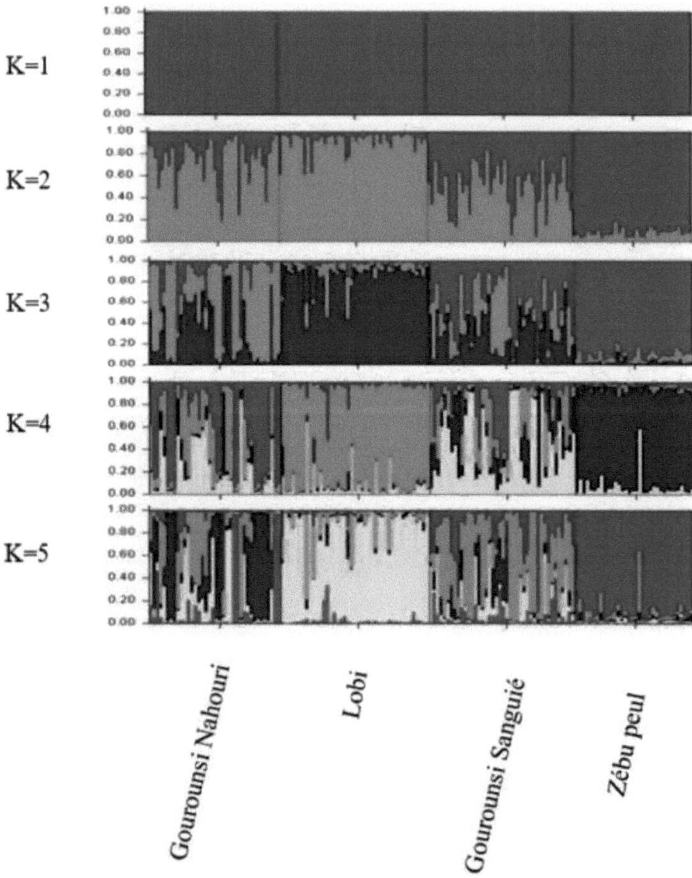

Figure 24: Genetic structure of populations at K=2

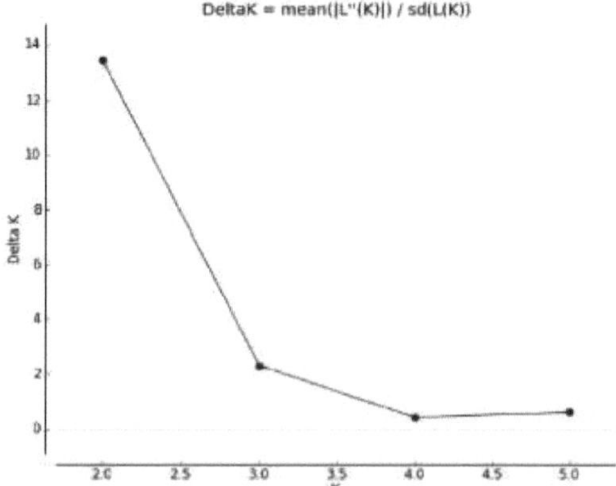

Figure 25: Genetic assignment of individuals, average probabilities of positivity (L(K))

The dendrogram obtained from the matrix of distances of shared alleles between populations (POSA POP) shows that the populations ëtudiëes sëparent into two perfectly distinct clades with a boostrap value of 100% (figure 26). The first clade groups the Gourounsi taurine populations, while the second clade is that of the lobi taurines.

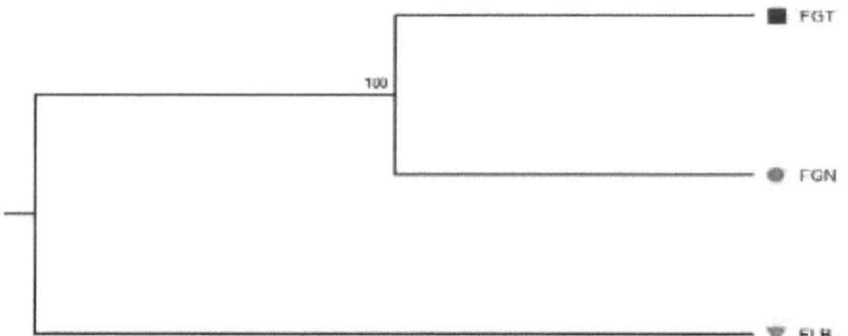

Figure 26: Phylogenetic tree inferred from the distance matrix of shared alleles by the N.J algorithm showing genetic relationships between populations. The nodes are supported by bootstrap percentages obtained after 10,000 replications.

-FGN: Gourounsi Nahouri, -FGT: Gourounsi Sanguie; -FLB: Taurins Lobi

The tree of individuals highlights the strong homogënëitë of individuals of the Lobi race, whose individuals group together in a homogëne cluster. In contrast, the clusters formed by Gourounsi taurines are not very homogeneous and consist of alternating successions of individuals belonging to the two Gourounsi Nahouri and Gourounsi Sanguië populations. However, individuals from the Gourounsi Nahouri subpopulation form clusters that overlap with lobi taurins and Gourounsi Sanguië (Figure 27).

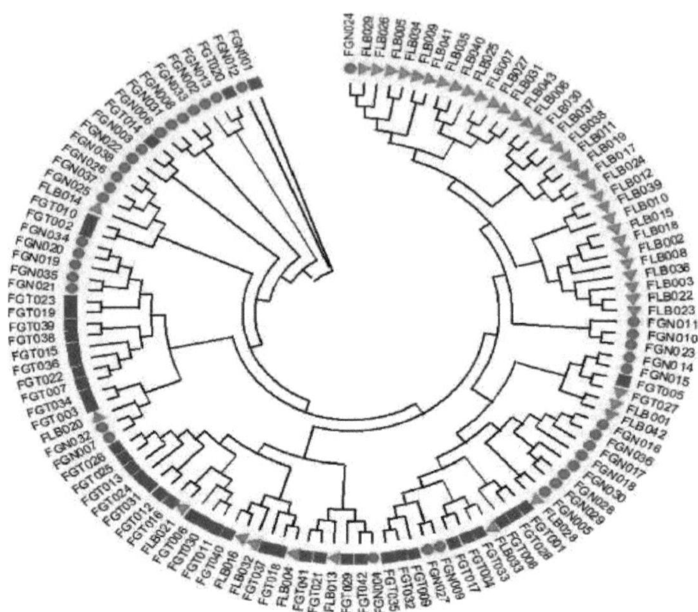

Figure 27: Neighbor-joining tree using distance matrices based on inter-individual shared alleles.

Individuals were assigned to their populations of origin using the Bayësian approach and likelihood methods. Regardless of the classification mëthod used, individuals belonging to the Lobi taurine population ële perfectly classifiedë in their populations of origin. On the other hand, in the Gourounsi bull populations, the classification results differed from one method to another, and from one population to another. Thus, the Gourounsi taurins of Nahouri (89 to 94.74%) had better scores compared with the Gourounsi Sanguië (47 to 97.67%) (Table XVI).

Table XV: Percentages (%) of individuals correctly assigned to gënëtic clusters using different methods.

Ramnala et		**Baudouin and Mountain (1997)**		**Lebrun (2001)**		**Paetkau *et al.* (1995)**	
Breeds	**N**	No. C.A.	SALES	No. C.A.	SALES	No. C.A.	SALES
GourN	38	34	89	36	94,74	36	94,74
GourS	43	20	47	42	97,67	41	95,35
Lobi	42	42	100	42	100	42	100
Total	123	96	78,05	120	97,56	119	96,75

No: Number; -N: Numbers, C.A.: Correctly Assigned

When a population faces a recent bottleneck, there is a significant reduction in the effective population size and thus the reduction in the number of alleles is faster than the reduction in the number of heterozygotes. Three mutation models, namely Infinite Allele Model (IAM), Two Phase Model of mutation (TPM) and Stepwise Mutation Model (SMM) were used in the BOTTLENECK programme to detect the existence of genetic bottlenecks using three tests (sign test, standard difference test and the wilcoxon sign rank test). The probability values obtained under these three models using the different statistical tests are presented in Table XVII. The results of the sign test showed that 23, 21 and 18 loci respectively showed excess heterozygosity under the IAM model in the Nahouri du Sanguie and Sud-Ouest (lobi) taurine populations (Table XVII). The expected number of loci with an excess of heterozygosity under the IAM mutation model was 15.98 for the Gourounsi taurines of Nahouri, 15.70 for the Lobi and 15.97 for the Gourounsi Sanguie taurines. However, only the values for the Gourounsi bull populations were significant (P <0.01). Thus, under the IAM mutation model of the sign test, the Gourounsi bulls would deviate from the mutation-derivative equilibrium. However, pairwise comparisons (observed heterozygosity excess vs expected heterozygosity excess) showed that under TPM, only the Gourounsi taurine populations showed significant heterozygosity excess whereas no deviation from the derivative mutation equilibrium was found under the SMM mutation model for the three populations. Standard differences tests showed negative T2 values for all three breeds under the different mutation models except for Gourounsi Nahouri under TPM and all populations under IAM. Similarly, the one-sided Wilcoxon sign rank test for excess genetic diversity revealed a significant deviation (P <0.01) for all breeds under IAM, whereas no deviation from the mutation-derivation equilibrium was observed under TPM and SMM in all populations except for Gourounsi Nahouri cattle under TPM.

Table XVI: Analysis of recent population bottleneck signatures using the Wilcoxon test.

Test	Parameters	Gourounsi Nahouri			Lobi			Gourounsi Sanguib		
		IAM	TPM	SMM	IAM	TPM	SMM	IAM	TPM	SMM
Sign test	Expected no, of loci with He excess	15,98	15,95	16,05	15,7	16,04	16,07	15,97	16,06	15,89
	Observed no, of loci with He excess	23	20	8	18	10	2	21	17	5
	P value	0,003	0,079	0,001	0,243	0,015	0	0,035	0,436	0
Standardized differences test	T2 value	3,032	0,61	-4,145	0,864	-2,604	-9,131	2,686	-0,343	-6,392
	P value	0,001	0,27	0	0,193	0,004	0	0,003	0,365	0
Wilcoxon sign rank test	P value (one tail for He excess)	0,001	0,027	0,998	0,097	0,984	1	0	0,365	0,999

The p-values correspond to one-sided Wilcoxon tests showing a significant excess of heterozygosity; S*: $p < 0.0001$ = highly significant; NS: not significant.*

In addition, the qualitative graphical mëthode basedëe on the frequency of rare allies showed a normal L-shaped curve, in the three populations of Taurine cattle which refutes that the population has not undergone any recent reduction in size (Figure 28).

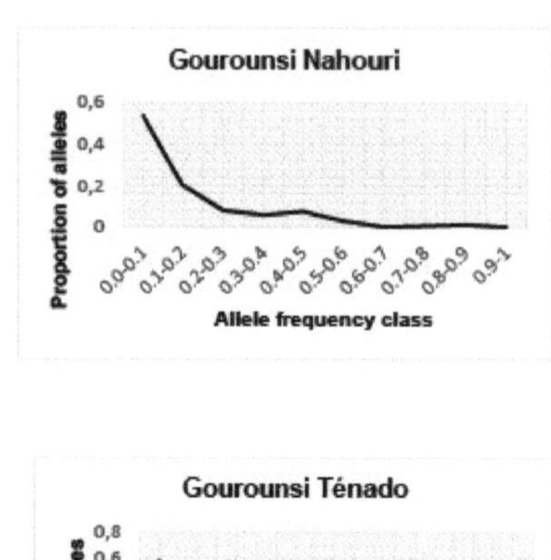

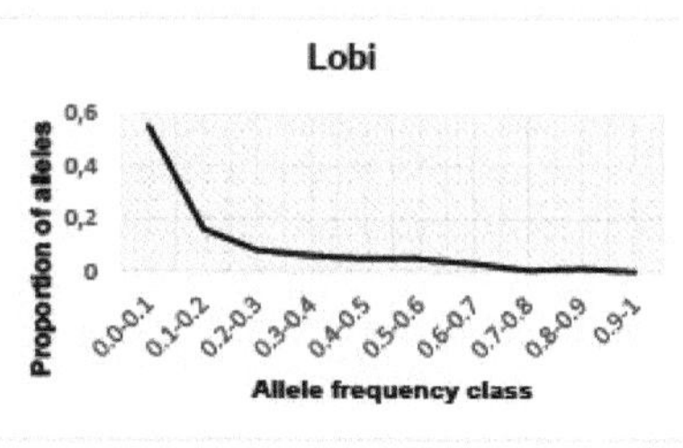

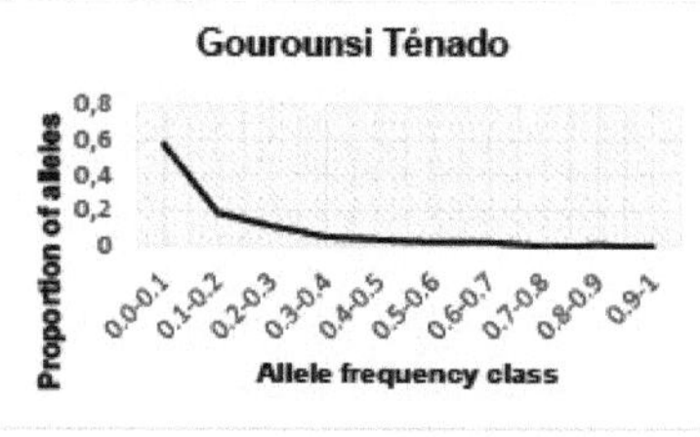

Figure 28: Results of Shift Mode Analysis showing the L distribution of allelic frequencies in the (a) Gourounsi Nahouri, (b) Lobi and (c) Gourounsi Sanguie bull populations.

III.3. Results of molecular characterisation of West African cattle.

The basic indices of gënëtic diversity are prësentës in Table XVIII. A total of 20520 genotypes were generated across the 27 STR loci used in this study. The mean number of observed alleles (Na) ranged from 3.7 (N'Dama) to 7.67 (Pure Arab) (Table XVIII). The mean observed and expected heterozygosity ranged from (0.508 and 0.715) to (0.577 and 0.748) respectively. The highest overall mean observed heterozygosity (breeds grouped by region and type) was observed in West African crossbred populations (0.704) when the lowest overall mean Ho value was reported in Asian zebu populations (0.597). A similar pattern was observed for the expected mean heterozygosity; West African crosses had the highest value (0.727) and the lowest value was found in European bull populations (0.627). The estimated heterozygosity deficit (ESD) ranged from -0.004 (Bororo Burkina) to 0.096 (Goudali Burkina) in West African zebu populations, while in Pakistan it ranged from 0.124 (Red Sindhi) to 0.060 (Tharparkar). In West African bull populations, FIS values ranged from 0.009 (Gourounsi Sanguie) to 0.094 (Lagunaire), while in European bull populations, the estimated heterozygosity deficit varied from 0.000 (Ayreshire) to 0.067 (Fleckvieh / Simmental).

Table XVII: Summary of genetic variability parameters for West African, European and Asian cattle populations

Type	Country	Breeds	Race code	N	Na	Ho	He	FIS	No. Loci Deviant de HWE	
									He Deficit	He Exces
Africa from l'Ouest -	Benin	Zebu Peuhl	EZP	34	7	0,67	0,6 99	0,04	4	0
	Burkina Faso Faso	Bororo Burkina Faso	FBO	20	6,3 7	0,70	0,7 0	0,00	2	1

Type	Country	Breeds	Code Breed	N	Na	Ho	He	FIS	No. Loci Deviant de HWE He Deficit	No. Loci Deviant de HWE He Excess
	Burkina Faso Faso	Goudali	FGO	37	6,96	0,65	0,71	0,10	7	0
	Niger	Pure Arabic	NAR	52	7,67	0,71	0,73	0,02	6	2
	Niger	Bororo Diffa	NBD	29	6,56	0,70	0,71	0,02	2	0
	Niger	Bororo site Kouri	NBK	50	6,96	0,66	0,69	0,04	6	1
	Niger	Bororo Maaoua	NBM	30	6,33	0,64	0,66	0,03	2	0
Crosses d'Afriq	Benin	Borgou x Zebu	EBZ	23	7,19	0,69	0,73	0,06	3	0
West	Niger	Kouri x Arabic	NKA	29	6,74	0,71	0,73	0,03	6	1
	Niger	Kouri x Bororo	NKB	40	7,22	0,72	0,72	0,01	3	4
Bullfighters d'Afriq	Benin	Borgou	EBG	14	6,07	0,70	0,75	0,07	7	0
West	Niger	Kouri Pur	NKO	36	6,85	0,69	0,72	0,05	5	1
	Benin	Lagoon	ELN	23	5,33	0,54	0,60	0,09	6	0
	Burkina Faso Faso	Gourounssi Nahouri	FGN	40	6,59	0,64	0,67	0,05	4	2
	Burkina Faso Faso	Gourounssi Bloody	FGT	39	7,3	0,70	0,71	0,01	3	0
	Burkina Faso Faso	Lobi Bouroum	FLB	41	6	0,58	0,60	0,03	2	1
	Mali	N'Dama	MND	8	3,7	0,51	0,5 8	0,13	3	0
Bullfighters Europee	India	Jersey Purebred	IJR	33	4,56	0,62	0,63	0,01	2	1
	India Sri Lanka	HF Purebred	ILH	53	6,85	0,68	0,72	0,05	6	1
	Sri Lanka	Ayreshire	LAY	27	5,19	0,68	0,68	0,00	2	1
	Austria	Fleckvieh/Simmental	AFS	42	5,81	0,61	0,65	0,07	5	0
Zebu Asian	Pakistan	Red Sindhi	PRS	30	6,96	0,61	0,69	0,12	6	0
es	Pakistan	Tharparkar	PTP	30	5,67	0,59	0,62	0,06	7	1

The F-statistics are shown in Table XIX. The mean estimate of the inbreeding coefficient (FIS) for all populations combined is 0.040±0.010 and ranges from -0.015±0.031 (SPS115) to 0.233±0.042 (INRA035). Of the 27 loci studied, 4 showed an excess of heterozygotes (ILSTS006, SPS115, INRA023 and TGLA122). Three (03) of these same loci (ILSTS006, SPS115, INRA023) also showed an excess of heterozygotes in African cattle populations. In addition, 3 other loci showed an excess of heterozygotes in African populations: CSRM60, BM1824 and TGLA126. Genetic differentiation attributed to breed differences (all populations combined) was 11.6% whereas it was 5.4% in West African populations.

Table XVIII: Overall F-statistics (mean ± standard deviation)

	All breeds (African, European and Asian)			**West African cattle**		
Locus	FIT	FST	FIS	FIT	FST	FIS
CSRM60	0,108±0,030	0,100±0,029	0,010±0,027	0,038±0,027	0,043±0,016	-0,005±0,034
CSSM66	0,141±0,040	0,119±0,036	0,024±0,020	0,072±0,022	0,040±0,014	0,033±0,020
HEL1	0,195±0,042	0,172±0,036	0,028±0,025	0,085±0,036	0,060±0,018	0,027±0,030
INRA63	0,199±0,041	0,177±0,042	0,027±0,024	0,094±0,040	0,043±0,021	0,053±0,030
BM1824	0,080±0,031	0,068±0,014	0,013±0,028	0,023±0,035	0,038±0,013	-0,016±0,035
ETH152	0,211±0,043	0,197±0,037	0,019±0,039	0,134±0,051	0,099±0,033	0,039±0,050
HAUT27	0,163±0,031	0,074±0,020	0,096±0,032	0,122±0,037	0,034±0,010	0,091±0,040
INRA05	0,068±0,028	0,065±0,026	0,003±0,025	0,032±0,026	0,021±0,007	0,011±0,029
BM1818	0,128±0,026	0,089±0,023	0,043±0,020	0,084±0,025	0,042±0,010	0,043±0,025
ETH3	0,145±0,040	0,121±0,038	0,028±0,023	0,111±0,033	0,079±0,032	0,035±0,023
HEL9	0,089±0,027	0,075±0,021	0,015±0,024	0,060±0,026	0,048±0,015	0,013±0,025
ILSTS006	0,082±0,052	0,086±0,023	-0,005±0,038	-0,012±0,041	0,036±0,013	-0,050±0,036
HAUT24	0,160±0,029	0,103±0,022	0,064±0,028	0,103±0,020	0,045±0,016	0,061±0,024
	All breeds (African, European and Asian)			**West African cattle**		
Locus	FIT	FST	FIS	FIT	FST	FIS
HEL5	0,288±0,038	0,123±0,034	0,188±0,036	0,230±0,032	0,043±0,013	0,196±0,041
INRA032	0,174±0,044	0,156±0,039	0,021±0,024	0,077±0,034	0,063±0,021	0,015±0,025
SPS115	0,150±0,085	0,162±0,076	-0,015±0,031	-0,010±0,027	0,034±0,011	-0,046±0,028
ETH185	0,192±0,029	0,109±0,025	0,093±0,024	0,166±0,038	0,058±0,024	0,114±0,029
HEL13	0,244±0,040	0,192±0,039	0,064±0,023	0,151±0,034	0,082±0,025	0,075±0,026
ILSTS05	0,117±0,033	0,086±0,021	0,033±0,027	0,095±0,041	0,046±0,016	0,051±0,034
INRA035	0,310±0,052	0,099±0,035	0,233±0,042	0,193±0,040	0,036±0,016	0,164±0,038
TGLA126	0,092±0,026	0,089±0,018	0,003±0,018	0,046±0,027	0,052±0,017	-0,006±0,020
BM2113	0,149±0,042	0,107±0,025	0,047±0,028	0,071±0,029	0,068±0,016	0,003±0,022
ETH10	0,151±0,030	0,133±0,022	0,021±0,018	0,097±0,031	0,092±0,023	0,005±0,021
ETH225	0,143±0,034	0,127±0,025	0,019±0,020	0,072±0,040	0,060±0,022	0,012±0,026
INRA023	0,125±0,038	0,129±0,028	-0,005±0,022	0,057±0,035	0,081±0,025	-0,026±0,024
TGLA122	0,116±0,037	0,116±0,028	-0,001±0,018	0,068±0,027	0,056±0,015	0,012±0,019
Total	0,151±0,011	0,116±0,007	0,040±0,010	0,087±0,010	0,054±0,004	0,034±0,011

The values of Nei distances between pairs of populations range from 0.033 (Zebu Peuhl-Borgou) to 0.311 (Jersey Purebred-Tharparkar). The gënëtic distance of Nei per pair of populations varies between 0.047 (Benin Zebu Peuhl - Bororo site Kouri) and 0.595 (Fleckvieh / Simmental-Tharparkar) (table XX).

TableXIX: FST by population pairs (top triangle) and Nei gënëtic distances (bottom triangle).

	PRS	PTP	EZP	FBO	FGO	NAR	NBD	NBK	NBM	EBZ	NKA	NKB	EBG	ELN	FGN	FGT	FLB	AND	NKO	IJR	ILH	LAY	AFS
PRS	PRS	-	0,06	0,095	0,086	0,07	0,077	0,081	0,12	0,12	0,096	0,078	0,099	0,085	0,227	0,177	0,133	0,241	0,243	0,099	0,257	0,208	0,221
PTP	PTP	0,124	-	0,151	0,141	0,12	0,123	0,126	0,162	0,175	0,145	0,127	0,148	0,142	0,285	0,229	0,182	0,284	0,305	0,146	0,311	0,258	0,271
EZP	EZP	0,183	0,239	-	0,008	0,019	0,02	0,022	0,013	0,012	0,013	0,018	0,015	0,009	0,139	0,081	0,048	0,116	0,146	0,036	0,19	0,157	0,177
FBO	FBO	0,201	0,251	0,066	-	0,01	0,013	0,02	0,013	0,02	0,012	0,018	0,022	0,015	0,149	0,084	0,045	0,125	0,157	0,033	0,198	0,157	0,179
FGO	FGO	0,164	0,215	0,059	0,073	-	0,011	0,032	0,034	0,039	0,011	0,017	0,027	0,011	0,151	0,093	0,051	0,132	0,158	0,035	0,189	0,154	0,172
NAR	NAR	0,182	0,229	0,063	0,077	0,05	-	0,017	0,037	0,046	0,01	0,011	0,014	0,017	0,122	0,065	0,035	0,11	0,128	0,018	0,173	0,146	0,152
NBD	NBD	0,19	0,236	0,068	0,089	0,085	0,059	-	0,042	0,045	0,022	0,013	0,02	0,023	0,138	0,068	0,045	0,116	0,14	0,027	0,195	0,15	0,164
NBK	NBK	0,207	0,241	0,047	0,067	0,066	0,069	0,086	-	0,016	0,024	0,037	0,029	0,026	0,139	0,092	0,058	0,113	0,142	0,053	0,188	0,161	0,184
NBM	NBM	0,209	0,248	0,058	0,086	0,085	0,103	0,105	0,053	-	0,043	0,04	0,033	0,039	0,163	0,11	0,066	0,137	0,173	0,058	0,209	0,177	0,199
EBZ	EBZ	0,219	0,255	0,064	0,083	0,068	0,069	0,089	0,069	0,102	-	0,012	0,01	0,003	0,102	0,046	0,025	0,079	0,112	0,015	0,168	0,139	0,157
NKA	NKA	0,19	0,242	0,074	0,089	0,069	0,062	0,073	0,082	0,102	0,076	-	0,015	0,015	0,133	0,074	0,041	0,11	0,139	0,022	0,186	0,143	0,162
NKB	NKB	0,205	0,252	0,061	0,092	0,071	0,061	0,073	0,07	0,085	0,07	0,061	-	0,019	0,126	0,059	0,036	0,092	0,121	0,011	0,176	0,142	0,16
EBG	EBG	0,22	0,267	0,085	0,106	0,093	0,105	0,111	0,087	0,112	0,091	0,099	0,089	-	0,115	0,062	0,027	0,103	0,119	0,024	0,168	0,125	0,152
ELN	ELN	0,393	0,452	0,224	0,237	0,241	0,209	0,227	0,217	0,258	0,204	0,235	0,223	0,222	-	0,053	0,065	0,068	0,092	0,117	0,205	0,165	0,189
FGN	FGN	0,311	0,374	0,14	0,137	0,154	0,126	0,128	0,146	0,177	0,117	0,143	0,127	0,151	0,132	-	0,022	0,033	0,058	0,053	0,154	0,13	0,143
FGT	FGT	0,245	0,291	0,093	0,098	0,098	0,081	0,097	0,102	0,126	0,085	0,098	0,093	0,11	0,142	0,067	-	0,05	0,074	0,032	0,148	0,119	0,133
FLB	FLB	0,391	0,436	0,181	0,189	0,2	0,173	0,181	0,181	0,214	0,15	0,177	0,159	0,193	0,118	0,078	0,087	-	0,063	0,094	0,196	0,168	0,189
MND	MND	0,461	0,514	0,267	0,277	0,283	0,251	0,255	0,257	0,294	0,244	0,26	0,241	0,256	0,178	0,153	0,174	0,133	-	0,107	0,211	0,177	0,193
NKO	NKO	0,214	0,253	0,087	0,103	0,082	0,063	0,078	0,095	0,115	0,074	0,071	0,048	0,104	0,211	0,113	0,089	0,154	0,22	-	0,177	0,144	0,152
IJR	IJR	0,533	0,584	0,364	0,376	0,38	0,349	0,373	0,363	0,393	0,341	0,384	0,355	0,38	0,317	0,275	0,295	0,279	0,356	0,353	-	0,095	0,133
ILH	ILH	0,469	0,54	0,338	0,356	0,339	0,327	0,336	0,343	0,366	0,329	0,333	0,31	0,332	0,297	0,269	0,268	0,273	0,355	0,311	0,185	-	0,066
LAY	LAY	0,505	0,564	0,38	0,398	0,369	0,347	0,37	0,386	0,409	0,369	0,368	0,356	0,378	0,326	0,295	0,296	0,3	0,362	0,336	0,242	0,148	-
AFS	AFS	0,544	0,595	0,427	0,421	0,418	0,398	0,431	0,403	0,449	0,392	0,415	0,403	0,416	0,397	0,361	0,364	0,374	0,459	0,394	0,309	0,259	0,328

The results of the Analysis of Molecular Variance (AMOVA) are prësentës in Table XXI. All populations groupedëes; the main source of variation is attributed to differences within populations (88.41%) when the variability between populations was 11.59%. When populations were grouped according to their genetic type (*Bos indicus vs Bos taurus*), variation within populations was estimated at 87.04% while variation between groups collapsed to 7.63%. The remaining 5.34% of variability is attributed to differences between populations within groups. A similar level of variability is found when populations are grouped according to breed type, origin and purity.

Table XX: Results of the analysis of molecular variance

Groups	**Source of**	**d.l**	**Sum of**	**Variance**	**Percentage**	**P**

	variation		edges	composition	change	
No grouping	Between populations	22	1284,00	0,79	11,59	0,000
	In populations	1497	9083,41	6,06	88,41	0,000
Grouping I (By type of cattle)	Between groups	4	740,83	0,53	7,63	0,000
	Between populations within groups	18	543,17	0,37	5,34	0,000
	In populations	1497	9083,41	6,06	87,04	0,000
Group II (Group I - PRS, PTP (Pakistani zebu); Group II - EZP, FBO, FGO, NAR, NBD, NBK, NBM (African zebu); Group III - EBZ, NKA, NKB (African Crusaders); Group IV - ELN, FGN, FGT, FLB, MND (African bullfighters); Group V - IJR, ILH, LAY, AFS (European bullfighters);	Between groups	5	795,51	0,54	7,79	0,000
	Between populations within groups	17	488,49	0,34	5,02	0,000
	In populations	1497	9083,41	6,06	87,20	0,000

Groups	Source of variation	d.l	Sum of carres	Variance composition	Percentage change	P
Group VI - NKO, EBG (pure African bulls)						
Grouping III (Group I - PRS, PTP (zebu Pakistani); Group II - EZP, FBO, FGO, NAR, NBD, NBK, NBM, NKO, EBG (zebu and taurines Group III - EBZ, NKA, NKB (African crossbreeders); Group IV - ELN, FGN, FGT, FLB, MND (African bullfighters); Group V - IJR, ILH, LAY, AFS (European bullfighters).	Between groups	4	771,51	0,57	8,27	0,000
	Between populations within groups	18	512,49	0,34	4,95	0,000
	In populations	1497	9083,41	6,06	86,79	0,000

-dl: degree of НЬеЛё

The classic sclk'ina of West African zëbu descended from Asian *B. indicus* is respected in the NJ tree (Figure 29a). In accordance with this first observation, European cattle breeds and certain African bull breeds form a clade sëparë. This clade is in an intermediary position between the clades formedë by supposedly pure zëbus and croisës. Surprisingly, the Kuri taurine shares the same clade with the mëtis from the Bororo x kuri cross. The Borgou (EBG) from the Bënin shares the pure zëbus clade with a high bootstrap value (73%). This suggests the likely introgression and diffërence in introgression level of kuri and EBG *B. indicus*. Furthermore, despite the existence of croisës (Zebu-taurines) in our ëtude, it is intëressant to note that not all mëtis cluster together. In fact, EZB (Zëbu peul x Borgou) and NKA (Kouri x Arabe) were found to be closer to zëbu strains than NKB (Kouri x Bororo). This observation reveals the existence of different levels of introgression in the various crossbred populations.

The еигорёеппе and Pakistani breeds group together and form sëparës clades. African taurines also form ë a sëparë clade, although breeds such as FGN (Gourounsi Nahouri), FGT (Gourounsi Sanguië) and EBG (Borgou) are close to the Юпиё clade by African zëbus. Furthermore, the Kouri breed shares the same clade as the croisës (Figure 29b).

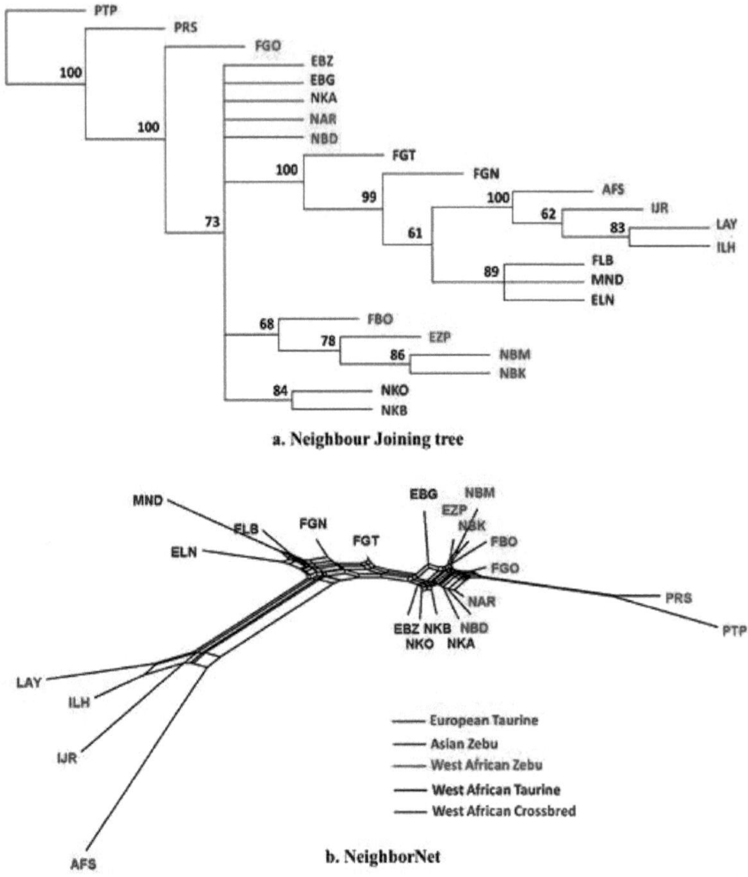

Figure 29: (a) Phylogenetic tree inferred from the distance matrix of Nei shared by the N.J algorithm showing the genetic relationships between populations. The nodes are supported by bootstrap percentages obtained after 10,000 replications. (b) N.J network inferred from the Nei sharing distance matrix

The first dimension makes it possible to clearly distinguish between bulls on the one hand, and zëbus and croisës on the other. The second dimension makes it possible to sëparate African bull breeds and mëtisses on the one hand. The second dimension also makes it possible to sëparate the bull breeds of West Africa and those of Europe (Figure 30).

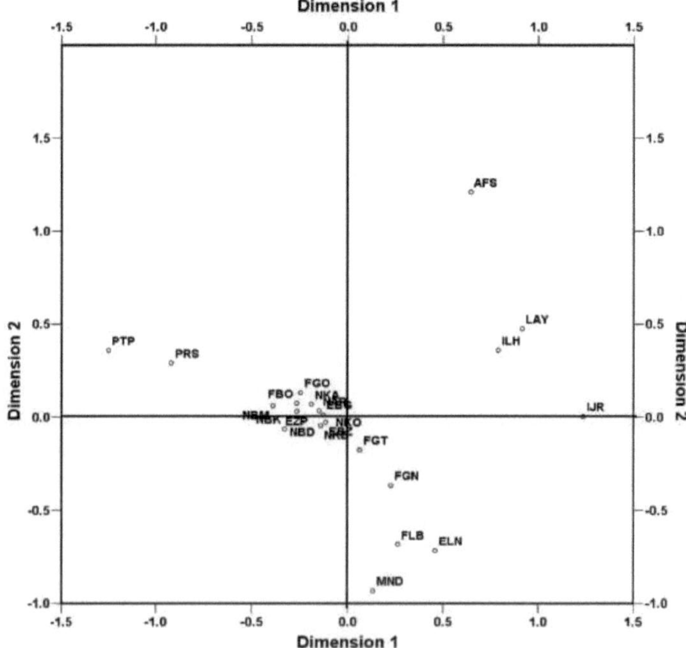

Figure 30: Multidimensional representation of FST per pair in zebu, bull and crossbred cattle in West Africa.

The European taurines (CET) and the Asian zëbus (ASZ) represent two different and well-separated appendages on the scattergram; while the West African taurine populations, the metis and the zebus are in an intermediate position (Figure 31). However, the kuri, a supposedly pure bull breed, overlaps with the group formed by the West African zebus. As with some taurine populations, crossbreeds also overlap with clusters formed by zebus and West African taurines.

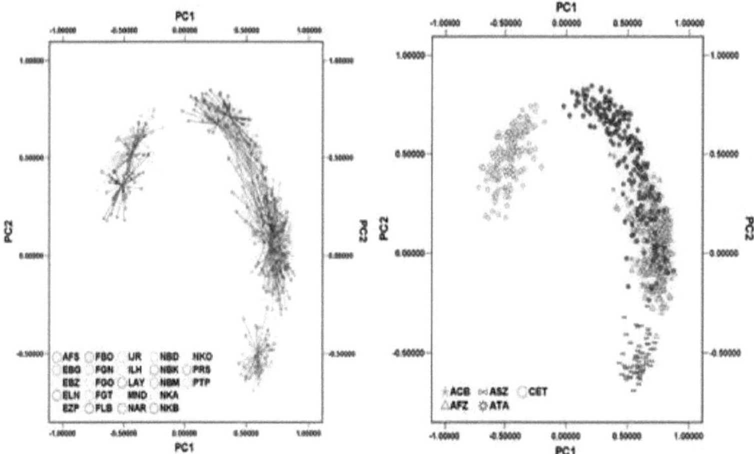

Figure 31: Scattergram of the first and second largest principal components derived from the inter-individual shared allele distance by population pairs between West African zebu, taurine and crossbred cattle. (ASZ-West African Zebu; ATA-West African Taurine; ACB-West African Metis)

Assuming that the most likely number of ancestors is 2, the populations have been groupedëes into 2 groups. The European taurine populations were clearly distinguished from all the African populations. These were identified as sharing gënëtic material with Asian zëbus (Figure 32). At k = 3, all African bull populations are different from their European counterparts. However, the Kuri ëtaient complement introgressës par le zebu (90%). At this level, no clustering appeared between African zebu, African metis and Asian zebu. At k = 4, the difference between African and Asian zebu is clearly made. Although African zebus and metis still group together. With the exception of the N'dama, Lobi Bouroum and Lagunaire populations, all the supposedly pure African bull breeds are inbred with the zebu genotype. The Gourounsi Sanguie bulls show the highest level of introgression, which is around 30%. K = 5 seems to be the most likely number of common ancestors to clearly distinguish the populations studied. In this respect, all the populations are clearly separated and their genotypes well represented. Surprisingly at this stage, all African zebu are introgressed and populations such as Bororo Diffa and Arabe Pure showed a level of introgression similar to those found in the Kuri x Arab and Kuri x Bororo cross population (around 80%).

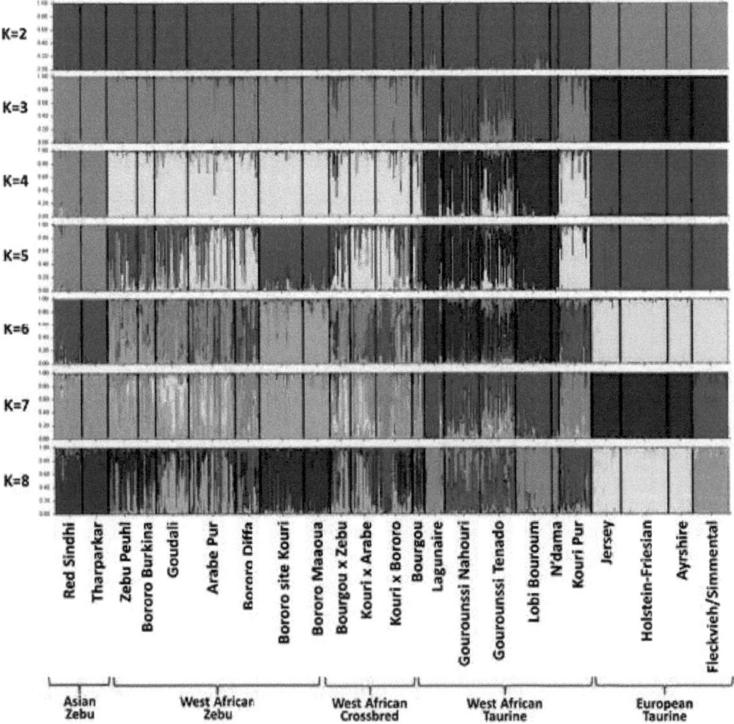

Figure 32: Bayesian grouping of 760 cattle under the hypothesis of 2 to 8 clusters with no prior information on the populations.

CHAPTER IV

DISCUSSION

IV.1 Morpho-biometric characterisation

Morphological measurements are an excellent means of inferring gënëtic variability within and between animal populations. A great deal of characterisation work has indeed been undertaken on many animal species around the world to establish intra- and inter-population diversity (Ndumu *et al.*, 2008; Khan *et al.*, 2013; Boujenane and Petit, 2016; Traore *et al.*, 2016). On the basis of their morphological characteristics, the bull populations of Burkina Faso can be divided into 3 subgroups according to size. From the smallest to the largest, these are the Lobi, the GourN and the GourS. Similar observations were reported by Traore *et al* (2016). These observations corroborate the results obtained by Rege *et al.* (1994) on West African bull breeds. Changes in morphological parameters were observed both within and between populations. This evolution affected the distribution of variability between 2014 and 2018. Principal component analysis indicates that in 2014, bull populations in Burkina Faso were structured according to their morphology. The GourS formed a distinct population from the lobi and GourN.

Based on data from 2014, the GourS differed from the other bull populations in that it appeared to constitute a population in its own right, distinct from the other two bull populations. However, in 2018, the increase in measurements recorded in the Lobi and GourN taurine populations resulted in an overlap between the cluster circles of these two populations. This overlap reflects the weak structuring of bull populations in Burkina Faso in 2018. However, the ascending hierarchical classification indicated that: a) the cluster formed by the lobi is homogeneous, b) there is a homogenisation of the GourS and GourN which are grouped together in the same cluster. Comparing the results for 2014 and 2018, principal component analysis and ALD indicate a homogenisation of the GourS and GourN populations in 2018. The likely causes would be the existence of significant gene flow between these two populations or introgression by zebu into these two populations. Ibeagha-Awemu & Erhardt (2005), in a study of taurine and zebu populations in Africa, showed that neighbouring populations tend to differ little due to the existence of significant gene flow. Evidence of reproductive interchange between these two populations has been reported in the ALD. In fact, around 20% of GourS and GourN were reciprocally misclassified in the neighbouring population. This result highlights the existence of possible gene flow between these two populations. The province of Sanguie, where the GourS were sampled, is characterised by average rainfall which, as the dry season approaches, forces some herders to migrate westwards to Nahouri, where rainfall is more generous and pastures are more luxuriant. As in several West African countries, Burkina Faso is characterised by the absence of sëlections programmes and an extensive breeding system dominated by transhumance and panmictic mating on pasture (Traore *et al.*, 2016). Thus 1 the existence of reproductive exchanges between these populations on the pastures could explain their homogenisation.

However, in 2014, only 1.5% of GourNs were misclassified as GourSs, but between 2014 and 2018, no significant changes were made to the breeding system in Burkina Faso. Furthermore, the Gourounsi bulls (GourS and GourN) in 2014 had morphological averages that were clearly superior to those of the lobi. In 1987, when describing the Gourounsi bull populations, Planchenault referred to them as 'mere'. Mere is a local Fulani name for the metis, the result of cross-breeding between zebu and taurins. So the difference between Lobi taurines and Gourounsi taurines could be linked to the fact that the Gourounsis have been retrogressed by zebu genes. Consequently, the second hypothesis, according to which a homogenisation of Gourounsi taurine populations in 2018 would result from the introgression of zebu genes into these populations, is perfectly plausible and should also be taken into consideration. Consequently, and seen from this angle, the morphological differences reported between GourS and GourN could be explained by the difference in the level of zebu gene introgression in these two populations.

Studies on the level of introgression of African cattle breeds undertaken by Hanotte *et al* (2002) and Freeman *et al* (2006) revealed that the degree of introgression exhibited a west-east gradient.) Here, we provide evidence based on variation in body traits that would be consistent with this introgression model. MacHugh *et al* (1997) explain this situation by the increase in the presence of tsetse flies as one approaches western counterparts. Lobi taurines are mainly found in the Sudano-Guinean zone of Burkina Faso. This agro-ecological zone is characterised by a humid tropical climate with gallery forests dotted with numerous watercourses. This biotype is favourable to the development of trypanosomosis vector tsetse flies (Dayo *et al.*, 2010). Given the susceptibility of zebu to animal trypanosomosis, the distribution areas of lobi taurines are not easily accessible to zebu populations, which could explain the relatively better structure and homogenous character of lobi taurines between 2014 and 2018.

Cultural reasons must also be taken into account when interpreting the data. In Lobi country, a pure Lobi bull is required for the dowry ceremony. This helps to maintain a certain level of purity within the different herds. However, ongoing climate change and the existence of control programmes such as PATTEC have significantly reduced the density of tse-tse flies in the taurin lobi biotope. This would therefore have favoured the penetration of zebu into these sanctuaries. In addition, despite the natural barriers, some farmers force cross-breeding between zebu and taurin lobi. Ouedraogo *et al* (2020), trying to implement a community-based selection programme in the south-west, noted that farmers preferred large animals, which would be more interesting in terms of production and would be better suited to ploughing. These different elements would explain the existence of poorly classified Lobi bulls found in the different sub-populations of Gourounsi bulls and this in relation to the level of cross-breeding achieved.

IV.2 Molecular characterisation of Burkina taurins and West African cattle breeds

In the taurine populations of Burkina Faso, almost all the loci selected had a minimum of 4 alleles per locus. The loci chosen, with the exception of INERA035, all satisfy the rule of thumb given by the FAO, according to which markers for diversity studies must display at least 4 alleles per locus. According to this criterion, 26 of the 27 loci used in this study showed sufficient polymorphism to assess genetic variability in the different bull populations of Burkina Faso. The populations studied showed substantial genetic variability based on the average number of alleles per locus and the heterozygosity values observed. A similar level of diversity has been reported in cattle populations in Niger (Grema *et al.*, 2017; Moussa *et al.*, 2019), Senegal (Ndiaye *et al.*, 2015) and Ethiopia (Dadi *et al.*, 2008). However, these estimates are lower than those reported in Sudanese (Hussein *et al.*, 2015), Cameroonian (Ngono Ema *et al.*, 2014) and Central African (Ibeagha-Awamu *et al.*, 2004, Moazami Goudarzi et al., 2001) cattle populations. Out of 81 breed x locus combinations tested, 13.58% significantly deviated from the HWE. 72.7% of these deviations were due to a heterozygosity deficit. Deviations from panmictic equilibrium due to a heterozygosity deficit may be the result of a subdivision of the population (Waples, 2015), which is related to a sampling of apparent individuals. In addition, the Walund effect could be one of the possible causes of the observed deficit of heterozygotes, given that the samples were collected from different geographical locations. Similar levels of deviations from panmictic equilibrium have been reported in bull populations in Niger (14.8% breed x locus combinations) (Moustapha et al. (2017). With an estimated mean FST of 0.04, 4% of the total variation observed in bull populations in Burkina Faso is attributed to breed differences. The overall FST observed in Burkinabe bulls was well below that reported for Sudanese (FST = 0.084) (Hussein *et al.*, 2015) and Cameroonian (FST = 0.061) populations Ngono Ema *et al.* (2014). Furthermore, when considering pairwise FST per population, the highest genetic differentiation was reported between Lobi and Gourounsi sanguie (FsT = 0.05). This can be explained by the geographical isolation of these populations. In fact, Lobi cattle are found in the south-western part of the country, whereas Gourounsi cattle are found in the central part of Burkina Faso. In contrast, Gourounsi Tenado and Gourounsi Nahouri had the lowest FsT value. These cattle live in Gourounsi country and are neighbours. Ibeagha-Awemu and Erhardt (2005) pointed out that neighbouring domestic populations are generally poorly differentiated due to the existence of significant gene flow between them. This observation is reflected in the estimated gene flow (Nm) between pairs of populations. The proximity of Gourounsi cattle and their relatively low genetic differentiation compared with the Lobi were also observed on the population tree, where Gourounsi cattle populations are grouped together supported by a high bootstrap value. The weak structure of Gourounsi cattle and their differentiation from the Lobi may result from one or more of the following reasons: (i) reproductive isolation, (ii) introgression of zebu gënes (iii) differences in herd management. As pointed out earlier, the existence of geographical barriers between Lobi and Gourounsi taurines affects gene flow. The Lobi cattle are found in the wettest part of the country. This zone is typical of the tropical Sudanian climate, with bushy savannahs and gallery forests. These types of vegetation are ideal biotopes for the proliferation of tsetse flies (Dayo *et al.*, 2010). Given the sensitivity of zebu to trypanosomosis, the Lobi range is more difficult to access for zebu, which are trypano-susceptible. In contrast, Gourounsi cattle are found in the central region of the country where rainfall is average. However, the Gourounsi Nahouri live in a more humid area. Consequently, the differences between Lobi and Gourounsi cattle may be linked to the introgression of Zebu genes into the Gourounsi cattle populations. This latter hypothesis is supported by the results of sTRUCTURE.

In the present study, allelic frequencies were used to perform genotype assignment based on probabilities and Bayesian methods. The Lobi cattle were assigned 100% correct compared with the mixed Gourounsi cattle populations. These results reflect the weak structuring of Gourounsi cattle as shown in the previous results.

Compared with European and Asian populations, West African cattle show a high level of genetic variability, characterised by low inbreeding and high allelic variability. Some authors, such as Thevenon *et al* (2007), explain this high genetic diversity by the existence of a large effective population size resulting from the effects of extensive livestock farming marked by the absence of selection and dominated by pastoral practices such as nomadism or transhumance, which favour this variability. In addition, West African breeds generally have a relatively weaker genetic structure compared with European and Asian breeds. This structure is linked to the existence of significant gene flow between the different populations. In contrast, the trypanotolerant bull breeds show a stronger genetic structure. This observation can be attributed to the existence of strong pressure from tse-tse flies, which tend to maintain the genetic make-up of these populations intact. This is the case for the N'dama and the Lobi taurins. In addition, and as has been shown for certain bull breeds in Burkina Faso, cultural reasons must be taken into account to better understand these observations. In general, West African zebu breeds are characterised by low FST values and high gene flow between them relative to a high migration rate in a usual transhumant farming system (Freeman *et al.,* 2004; Dayo *et al.,* 2009).

GENERAL CONCLUSION AND OUTLOOK

Three bull populations from Burkina Faso were ëvaluëed for variation in body traits. A modële basedë on temporal variations in morphological measurements was usedë. The variability of body traits between and within populations changed considëraЬlement between 2014 and 2018. In 2014, the two Gourounsi bull populations (GourS and GourN) were clearly differentiated. In addition, a clear differentiation between Gourounsi and Lobi taurines was demonstrated; the Lobi were of smaller stature compared to the Gourounsi taurine populations. Within the breeds, the Gourounsi of Sanguie were larger than the Gourounsi of Nahouri. The differences within the Gourounsi taurines were probably due to differences in the level of introgression of zebu genes, GourS being more introgressed than GourN because of a lower density of tse-tse flies in the Sanguie region. However, a strong homogenisation signal was identified within the Gourounsi populations in 2018. Alongside introgression, there has also been an increase in gene flow between Gourounsi taurines through migration (transhumance). Between breeds, Lobi cattle seemed less different in 2018. Increased introgression of zebu due to climate change and farmers' preferences seems to have diluted the pure Lobi breed. However, cultural considerations and the relatively high pressure of trypanosomiasis in the Lobi breeding area have enabled a higher proportion of pure-bred cattle to be maintained. Current research is highlighting the rapid genetic erosion of bull populations in Burkina. This is consistent with all other African bull populations. These cattle possess a unique genetic assortment that deserved to be protected and fully explored. The lack of appropriate selection and conservation programmes for bulls is a major problem that needs to be addressed. Consequently, the evidence presented in this study can be used to save the taurine cattle populations of Burkina Faso. The aim is to put in place sustainable conservation and selection strategies within the bull populations of Burkina Faso. The implementation of sustainable selection programmes involves recording the production performance of the bull breeds in Burkina Faso through a longitudinal study. In addition, association studies could also be undertaken using SNP markers to identify QTLs in relation to animal performance, with a view to a genomic selection programme.

BIBLIOGRAPHICAL REFERENCES

Gatesy J., Yelon D., DeSalle R., Vrba E. S., 1992. Phylogeny of the *Bovidae* (Artiodactyla, Mammalia), based on mitochondrial ribosomal DNA sequences. *Molecular Biology and Evolution*, **9**(3):433-446.

Ajmone-Marsan P. Garcia. J.F. Lenstra. JA. 2010. On the origin of cattle: how aurochs became cattle and colonized the world. Evolutionary Anthropology: Issues, News, and Reviews. **19**(4): 148-157. https://doi.org/10.1002/evan.20267

Ajmone-Marsan P., Lenstra J. A., Kantanen J., and The European Cattle Genetic Diversity Consortium, 2007. Differentiation of European cattle by AFLP fingerprinting. *Animal Genetics*, 38: 60-66.

Alvarez I. Traore A. Fernandez I. Lecomte T. Soudre A. Kabore A. Tamboura H.H. Goyache F. 2014. Assessing introgression of Sahelian zebu genes into native Bostaurus breeds in Burkina Faso. Molecular Biology Reports, 41: 3745-3754. DOI 10.1007/s11033-014-3239-x

Alvarez I. Traore A. Tamboura H.H. Kabore A. Royo L.J., Fernandez I., Ouedraogo-Sanou G., Sawadogo L., Goyache F. 2009. Microsatellite analysis characterizes Burkina Faso as a genetic contact zone between Sahelian and Djallonkë sheep. *Animal Biotechnology*, 20: 47-57. https://doi.org/10.1080/10495390902786926

Anderson S., De Bruijn M.H., Coulson A.R., Eperon I.C., Sanger F. and Young I.G. 1982. Complete sequence of bovine mitochondrial DNA. Conserved features of the mammalian mitochondrial genome. *J. Mol. Biol.* 156: 683-717.

Arber W., Linn S., 1969. DNA modification and restriction. *Annual Review of Biochemistry*, 38**:** 467-500.

Arif I.A., Bakir M.A. and Khan H.A. 2012. Inferring the Phylogeny of Bovidae Using Mitochondrial DNA sequences: Resolving Power of Individuals Genes Relative to Complete Genomes. *Evol. Bioinform,* 8: 139-159.

Baker N. 2005. Levels and Patterns of Genectic Diversity in Wild Populations and Cultured stocks of Cherax quadricarinatus (von Martens, 1868) (Decapoda: Parastacidae). Doctoral thesis: B. App. Sci (Hons), Queensland University of Technology, 123p. Banik S. and Gandhi R.S. 2010. Sire evaluation using single and multiple trait animal models in Sahiwal cattle. *Indian. J. Anim. Sci.* 80: 269-270.

Baudouin L., Lebrun P., 2001. An operational bayesian approach for the identification of sexually reproduced cross-fertilized populations using molecular marker. In: DORE, C.; DOSBA, F.; BARIL, C. (Eds). *ISHS Acta Horticulturae 546* - International symposium on molecular markers for characterizing genotypes and identifying cultivars in horticulture. Montpellier, France: ISHS, 2001. p.81-94.

Belkhir K., Borsa P., Chikhi L., Raufaste N. and Bonhomme F. 2004. GENETIX 4.05, Windows TM software for population gënëtics. Laboratoire Gënome, Populations, Interactions, CNRS UMR 5171, Universite de Montpellier II, Montpellier (France). Available at: "http://Kimura.univ-montp2.fr/genetix/constr.htm#download",

-Benefice E., Barral H., Doudet G. 1993. Systemes de production d^levage au Sënëgal dans la region du Ferlo. *Memento de l'Agronome*, 4: 1183-1191.

Bitgood J.J., Somes R.G., 1993. Gene map of the chicken (Gallus gallus or G. domesticus). In: S.J. O'Brien (ed), Genetic maps, 4.332-4.342. Cold Spring Harbor Laboratory Press.

Blench R.M., MacDonald K.C. 2000. The origins and development of African livestock: archeology, genetics, linguistics and ethnography. New York: Routledge.

https://doi.org/10.4324/9780203984239
Boichard D., Le Roy P., Leveziel H. and Elsen J.M., 1998. Utilisation des marqueurs moteculaires en gënëtique animale. *INRA ProdAnim* ,11 :67-80.
-Boujenane I., Petit D. 2016. Between- and within-breed morphological variability in Moroccan sheepbreeds . *AnimalGeneticResources* , 58: 91-100.DOI:
https://doi.org/10.1017/S2078633616000059
- Bradley D.G., MacHugh D.E., Loftus R.T., Sow R.S., Hoste C.H., Cunningham E.P. 1994. Zebu-taurine variation in Y chromosomal DNA: a sensitive assay for genetic introgression in West African trypanotolerant cattle populations. *Animal Genetics*, 25 :7-12. https://doi.org/10.1111/j.1365-2052.1994.tb00048.x
Bruford M.W., Bradley D.G. and Luikart G. 2003. DNA markers reveal the complexity of livestock domestication. *Nature Reviews Genetics*, 4: 900-910.
Chambers G.K. and Macavoy E.S. 2000. Microsatellites: consensus and controversy. Comparative Biochemistry and Physiology Part B, Biochemistry and Molecular Biology, *Elsevier*, **126**(4): 487-494.
- Charrad M., Ghazzali N., Boiteau V., Niknafs A. 2014. NbClust: An R Package for Determining the Relevant Number of Clusters in a Data Set. *J. Stat. Soft,* 61:1-36. 328p. DOI: 10.18637/jss.v061.i06
Colville G. and Shaw T. 1950. Report of the Nigerian Livestock Mission to the Colonial Office. *HMSO*, London, 58p.
Cornuet J.M. and Luikart G. 1996. Description and power analysis of two tests for detecting recent population bottlenecks from allele frequency data. *Genetics*, 144:2001-2014.
Coyral-Castel S., Rame C., Fabre-Nys C., Monniaux D., Monget P., Fabre-Nys C., Monniaux D., Monget P. et Dupont J. 2009. Characterization of dairy cows carrying the "fertile+/+" or "fertile-/-"haplotypes for one QTL of female fertility located on chromosome 3. *Renc. Rech. Ruminants,* 16: 313-316.
Cushwa, W.T., Dodds K.G., Crawfer A.M. and Medrano J.F. 1996. Identification and genetic mapping of Random Amplified Polymorphic DNA (RAPD) markers to the sheep genome. *Mann. Genome*, 7: 580-585
- Dadi H., Tibbo M., Takahashi Y., Nomura K., Hanada H., Amano T. 2008. Microsatellite analysis reveals high genetic diversity but low genetic structure in Ethiopian indigenous cattle populations. *Anim. Genet*. 39, 425-431.
Dayo G.K., Bengaly Z., Messad S., Bucheton B., Sidibe I., Cene B., Cuny G., Thevenon S. 2010. Prevalence and incidence of bovine trypanosomosis in an agro-pastoral area of southwesternBurkinaFaso . ResVetSci88 : 470-477.
http://dx.doi.org/10.1016/j.rvsc.2009.10.010
De Marchi M., Dalvit C., Targhetta C. and Cassandro M. 2006. Assessing genetic diversity in indigenous Veneto chicken breeds using AFLP markers. *Animal Genetics*, 37: 101-105.
De Meeus T., 2012. Initiation a la gënëtique des populations naturelles : Application aux parasites et a leurs vecteurs. Marseille, IRD Editions, *Collection Didactiques*. P:335.
De Meeus T., Bëati L., Delaye C., Aeschlimann A., Renaud F. 2002. Sex-biased genetic structure in the vector of Lyme disease, *Ixodes ricinus*. *Evolution,* 56: 1802-1807.
De Meeus T., Gueguan J-F. and Teriokhin A.T. 2009. MultiTest V.1.2, a program to binomially combine independent tests and performance comparison with other related methods on proportional data. *BMC Bioinformatics,* 10: 443. Available at:

"http://www.biomedcentral.com/1471-2105/10/443", (consulted on 15 June 2014).

De Meeus T., Koffi B.B., Barre N., De Garine-Wichatitsky M., Chevillon C. 2010. Swift sympatric adaptation of a species of cattle tick to a new deer host in New-Caledonia. *Infect. Genet. Evol.* 10: 976-983.

Dieringer D., and Schlotterer C. 2003. MICROSATELLITE ANALYZER (MSA): a platform independent analysis tool for large microsatellite data sets. *Mol. Ecol. Notes*, 3, 167-169.

Doutressoulle G., 1947. L'elevage en Afrique occidentale frangaise. Paris, Larose, 299 p.

Earl, Dent A. and vonHoldt, Bridgett M., 2012. STRUCTURE HARVESTER: a website and program for visualizing STRUCTURE output and implementing the Evanno method. *Conservation Genetics Resources* 4 (2) pp. 359-361. doi: 10.1007/s12686-011-9548-7

Edwards C.J., Bollongino R., Schen A., Chamberlain A., Tresset A. 2007. Mitochondrial DNA analysis shows a Near Eastern Neolithic origin for domestic cattle and no indication of domestication of European aurochs. *Proc. R. Soc. B.,* 274: 1377-1385.

Egito A. A., Fuck B.H., Mcmanus C., Paiva S. R., Albuquerque M.M., Santos S. A., Pinto De Abreu U. G., Da Silva J. A., De Souza S.F.T.P., Mariante A., 2007. Genetic variability of Pantaneiro horse using RAPD-PCR markers. *R. Bras. Zootec*, **36** (4) 799-806.

Epstein H. 1971. The origin of the domesticated animals of Africa. Africana Publ. Corp, New York, London, Munich, 1-719 pp.

-Evanno G., Regnaut S., Goudet J. 2005. Detecting the number of clusters of individuals using the software structure: a simulation study. *Mol. Ecol.* 14,2611-2620.

Falush D. Stephens M., Pritchard J., 2003. Inference of population structure using multilocus genotype data: Linked loci and correlated allele frequencies. *Genetics.* 164. 1567-87. 10.3410/f.1015548.197423.

FAO (Food and Agricultural Organisation of the United Nations). 2004. Secondary Guidelines for development of National Farm Animal Genetic Resources using reference microsatellites. http://dad.fao.org./en/refer/library/guideline/marker.pdf.

-FAO, 2011. Guidelines to Phenotypic characterization of Animal Genetic Resources. Rome, 142p.

Felsenstein J. 1985. Confidence limits on phylogenies: An approach using the bootstrap. *Evolution,* 39: 783-791.

Felsenstein J. 1993. PHYLIP: Phylogeny inference package, version 3.5. Department of Genetics, Washington University, Seattle, Washington

Fernandez-Tajes J. and Mëndez J. 2007.Identification of the Razor Clam Species *Ensis arcuatus*, *E. siliqua*, *E. directus*, *E. macha*, and *Solen marginatus* Using PCR-RFLP Analysis of the 5S rDNA Region. *J. Agric. Food Chem,* **55** (18) : 7278-7282. DOI : https://doi.org/10.1021/jf0709855.

Flores E.B., Maramba J.F., Aquino D.L., Abesamis A.F., Cruz A.F. and Cruz L.C. 2007. Evaluation of milk production performance of dairy buffaloes raised in various herds of the Philippine Carabao Center. *Ital. J. Anim. Sci.* 6: 295-298.

-Freeman A., Bradley D.G., Nagda S., Gibson J.P., Hanotte O. 2006. Combination of multiple microsatellite data sets to investigate genetic diversity and admixture of domestic cattle. *Anim. Genet.* 37:1-9. DOI: 10.1111/j.1365-2052.2005.01363.x

Gellin J., Grosclaude F., 1991. Analyse du дёпоте des espëces d^levage. Project d^tablissement de la carte gënëtique du porc et des bovins. *INRA Prod. Anim.* 4, 97-105.

Goudet, J. 2002. Fstat Vision (2.9.3.2): A Computer Program to Calculate F-Statistics. Journal of Heredity, 86, 485-486.

Bovidae Gray, 1821 in Pyle R (2019). ZooBank. Version 1.505. International Commission on Zoological Nomenclature. DOI : https://doi.org/10.15468/wkr0kn.

Grema M., Traore A., Issa M., Hamani M., Abdou M., Soudre A., Sanou M., Pichler R., Tamboura H.H., Alhassane Y., Periasamy K., 2017. Short tandem repeat (STR) based genetic diversity and relationship of indigenous Niger cattle. *Arch. Anim. Breed*, 60, 399-408, 2017 https://doi.org/10.5194/aab-60-399.

Grodzicker T, Williams J, Sharp PA, Sambrook J., 1974. Physical mapping of temperature sensitive mutations of adenoviruses. Cold Spring Harbor Symp Quant Biol, 39: 439-46.

Grosclaude F., Aupetit R., Lefebvre J., Meriaux J.C., 1990. Essai d'analyse des relations gënëtiques entre les races bovines frangaises a l'aide du polymorphisme biochimique. *Genet. Sel. Evol.* 22: 317-338.

Guinko S. 1984. Vëgëtation of Upper Volta. Thëse de doctorat es sciences. Universite de Bordeaux III. Tome I, pp. 394.

Guintard C., Mangin J-P., Lignereux Y., 2008. Origine et diversite des Bovines-Domestications et representations : l'exemple de la phi^lie. Confërences de l'Assembtee Generale du SIERDA. Pontarcher (Aisne), 2 February 2008, 1-31. Available at: "http://www.oniris-nantes.fr/fileadmin/redaction/Sierda/PDF/17_-_Guntard_et_al.pdf", (consulted on 18 March 2020).

Gwakiska, P.S., Kemp S.J. and Teale A.J., 1994. Characterization of Zebu cattle breeds in Tanzania using random amplified polymorphic DNA markers. *Ani. Genet.* 25: 89-94.

Han S-H., Park S., OH H-S., Kang G., Park B-Y., Ko M-S., Cho S-R., Kang Y-J., Kim S-G. and Cho I-C., 2013. PCR-RFLP for the Identification of Mammalian Livestock Animal Species. *J. Emb. Trans.* 28 (4): 355-360.

Hanotte Clutton-Brock J. 1989. Cattle in Ancient North Africa. In: The Walking Larder Patterns of Domestication, Pastoralism, and Predation (ed. Clutton-Brock J), pp. 200-214.

Hanotte O., Bradley D.G., Ochieng J.W., Verjee Y., Hill E.W., Rege J.E. 2002. African pastoralism: genetic imprints of origins and migrations. *Science*, 296:336-339. DOI: 10.1126/science.1069878

Hassanin A. and Douzery E.J.P., 2005. Molecular and Morphological Phylogenies of Ruminantia and the Alternative Position of the Moschidae. *Systematic Biology* **52**(2):206-228.

Hiendleder S., Levalski H., Junke A., 2008. Complete mitochondrial genome of *Bos taurus* and *Bos indicus* provide new insights into intra- species variation, taxonomy and domestication. *Cytogenet. Genome Res.* 120: 150-156.

Hussein I.H., Alam S.S., Makkawi A.A.A., Sid-Ahmed S.E.A, Abdoon A.S., Hassanane M.S., 2015. Genetic Diversity Between and Within Sudanese Zebu Cattle Breeds Using Microsatellite Markers. *Research in Genetics*, 135483, https://doi.org/10.5171/2015.135483.

Ibeagha-Awemu E., Erhardt G. 2005. Genetic structure and differentiation of 12 African Bosindicus and Bostaurus cattle breeds, inferred from protein and microsatellite polymorphisms. *J Anim Breed Genet* 122:12-20. doi: 10.1111/j.1439-0388.2004.00478.x.

Ibeagha-Awemu E.M., Jann O.C., Weimann C., Erhardt G., 2004. Genetic diversity, introgression and relationships among West/Central African cattle breeds. *Genet. Sel. Evol.* 36, 673-90.

IBM Corp. Released 2019. IBM SPSS Statistics for Windows, Version 26.0. Armonk, NY: IBM Corp.

INSD (Institut national de la statistique et de la dëmographie), 2019. Annuaire statistique, 366p.

INSD (Institut national de la statistique et de la dëmographie), 2016. The statistical yearbook of Burkina Faso 250p
Kalinowski S.T., Taper M.L. and Marshall T.C. 2007. Revising how the computer program CERVUS accommodates genotyping error increases success in paternity assignment. *Molecular Ecology*, 16: 1099-1106. doi :http://dx.doi.org/10.111/j.1365-294X.2007.03089.X
Kantanen J., Vilkko J., Elo K. and Tanila A.M., 1995. Random amplified polymorphic DNA in cattle and sheep: Application for detecting genetic variation. *Anim. Genet,* 25: 315-320
Kassambara A. 2015. Factoextra: EXtract and visualize the results of mulitivariate Data Analyses R package Version 1.0.3. Online http//www.sthda.com (Accessed 14 March 2020)
Khan M., Rahim I., Rueff H., Jalali S., Saleem M., Maselli D., Muhammad S., Wiesmann U. 2013. Morphological characterization of the Azikheli buffalo in Pakistan. *Animal Genetic Resources,* 52: 65-70. DOI: https://doi.org/10.1017/S2078633613000027
Khodaei Motlagh M., Roohani Z., Zare Shahne A. and Moradi M. 2013. Effects of age at calving, parity, year and season on reproductive performance of dairy cattle in Tehran and Qazvin Provinces. *Iran. Res. Opin. Anim. Vet. Sci.* 3: 337-342.
Kumar S., Stecher G., Li M., Knyaz C., Tamura K., 2018. MEGA X: Molecular Evolutionary Genetics Analysis across Computing Platforms. *Mol. Biol. Evol.* **35**(6):1547-1549 doi:10.1093/molbev/msy096
Kumari N., Singh L.B. and Kumar S., 2013. Molecular characterization of goats using random amplified polymorphic DNA. *Am. J. Anim. Vet. Sci.* 8: 4549.
Lakouetene C. E. T., 1999. Elevage periurbain: les pratiques d'amëlioration gënëtique. identification of diseases specific to dairy herds. Memoire UPB/IDR (Bobo Dioulasso), 130 p.
Larrat R., 1988. Manuel des agents techniques de l'elevage tropical. Collection manuels et d^levage, 5e Edition, 534 pages.
Lenstra J.A. and Bradley D.G., 1999. Systematics and Phylogeny of cattle (ed. R. Fries and A. Ruvinsky) CAB © International, 1-14.
Lenth R. 2019. Emmeans: Estimated Marginal Means, aka Least-Squares Means. 2019 R Package Version 1.3.4. Online: https://CRAN.R-project.org/package=emmeans. [(Accessed 19 May 2020)];
Lewontin R.C., Hubby J.L., 1966. A molecular approach to the study of genetic heterosigosity in natural populations. I. The number of alleles at different loci in Drosophila pseudooscura. Genetics 54, 546-595.
Lhoste P., 1991.Cattle genetic resources of west Africa, in: Hickman C.G. (Ed.), Cattle Genetic Resources, World Animal Science B: Disciplinary Approach, Elsevier, Amsterdam, pp. 73-88.
Linnaeus C. 1758. Species *Bos taurus*. *Systema Naturae,* Tomus I. Holnriae. Impensis Direct, LAURENT II SAL VII, **10** (1), 71p.
Loftus R.T., MacHugh D.E., Bradley D.G., Sharp P.M., Cunningham P., 1994. Evidence for two independent domestications of cattle, Proc. Natl. Acad. Sci.USA 91, 2757-2761.
Luikart G. and Cornuet J.M. 1998. Empirical evaluation of a test for identifying recently bottlenecked populations from allele frequency data. *Conservation Biology* **12**(1):228-237
MacDonald K.C., Hutton MacDonald R. 2000. The origins and development of domesticated animals in arid West Africa. In: The Origins and Development of African Livestock: Archaeology,Genetics, Linguistics and Ethnography (ed. MacDonald KC), pp. 127-162. UCL Press, London. https://doi.org/10.1002/oa.576
MacEachern S., Mc Ewan J. and Goddard M. 2009. Phylogenetic reconstruction and the

identification of ancient polymorphism in the Bovini tribe (Bovidae, Bovinae). *BMC Genet,* 10: 1471-2164.
MacHugh D.E., Larson G., Orlando L. 2017. Taming the Past: Ancient DNA and the Study of Animal Domestication. *Annual Review of Animal Biosciences.* 5:329-351. https://doi.org/10.1146/annurev-animal-022516-022747
MacHugh D.E., Shriver M.D., Loftus R.T., Cunningham P., Bradley D.G. 1997. Microsatellite DNA variation and the evolution, domestication and phylogeography of taurine and zebu cattle (*Bos taurus* and *Bos indicus*). *Genetics* 146 :1071-1086.
Mahmoudi B., esteghamat O., Sharhriyar A. and Babayev M.S. 2012. Genetic characterization and bottleneck analysis of Korbi Jobnub Khorasan goats by microsatellite markers. *J. Cell Mol. Ecol.* 10: 61-69.
Mahrous K.F., Ramadan H.A.I., Abdel-Aziem S. H., Abdel M. M. and Hemdan D. 2011. Genetic variations between camel breeds using microsatellite markers and RAPD techniques. *J. Appl. Biosci,* 39: 2626 - 2634.
Marshall F., Hildebrand E. 2002. Cattle before crops: the beginnings of Food production in Africa. Journal of World Prehistory,16, 99-143. DOI: 10.1023/A:1019954903395
Martin G.B., Williams J.G. and Tanksley S.D. 1991. Rapid identification of markers linked to a Pseudomonas resistance gene in tomato by using random primers and near-isogenic lines. *Proc. Natl. Acad. Sci.* 88: 2336-2340.
Mason I.L. 1951. The classification of West African livestock. Communi-cation technique No 7. Commonwealth Bureau of Animal Breedina and Genetics. Commonwealth Agricultural Bureaux, Farnham Royal (England).
Meselson M., Yuan R. 1968. DNA restriction enzyme from E. Coli. *Nature,* 217 : 1110-1114
Ministere des Ressources Animales (MRA) 2005. Document National/Initiative Elevage Pauvretë Croissance (IEPC).125p.
Moazami-Goudarzi K., Belemsaga D.M.A., Geriotti G., Liloc' D., Fagbohoun N.T., Kouagou I., Sidibe I., Codjia V., Crimella M.C., Grosclaude F., Toure S.M. 2001. Caractdrisation de la race bovine Somba a l'aide de marqueurs moldculaires. *Elev. Med. Vet. Pays Trop,* **54**(2): 129138. DOI: 10.19182/remvt.9791
Morammazi S., Torshizi Vaez R., Rouzbehan, Sayyadnejad M.B. 2007. Estimates of genetic parameters for production and traits in Khuzestan buffalos of. *Iran. Ital. j. Anim. Sci.* 6: 421424.
Moussa M.M.A., Grema M., Tapsoba A.R.S, Issa M., Amadou T., Marichatou H., Pichler R., Soudre A., Sanou M., Tamboura H.H., Yenikoye A., Periasamy K. 2019. Analysis of the genetic diversity of the Bororo (Wodaabd) cattle breed of Niger using microsatellite markers. *Int. J. Biol. Chem. Sci.* **13**(2): 1109-1126. DOI : http://ajol.info/index.php/ijbcs.
MRA (Ministere des Ressources Animales) ,Burkina Faso Burkina Faso, Ministere des Ressources Animales (MRA), 2012[a] . Statistiques du secteur de l'dlevage.-Ouagadougou: Dierection Gdndrale de la prdvision des statistiques et de l'd'economie de l'dlevage. 155p
MRA (Ministere des Ressources Animales) ,Burkina Faso, 2003. Rapport national sur l'etat des ressources gdndtiques animales au Burkina Faso. 74 pages.
MRA (Ministere des Ressources Animales) ,Burkina FasoBurkina Faso, Mini stere des Ressources Animales (MRA), 2012b. Contribution de l^levage a l^conomie et a la lutte contre la pauvretë, les dëterminants de son dëveloppement, 72 p.
Nianogo AJ, Sanfo R, Kondombo SD. Neya SB, 1996. Le point sur les ressources gënëtiques en тaйëre d'elevage au Burkina Faso. *Animal Genetic Resources Information* (AGRI),

FAO/UNEP, 13-31.
N'Goran K.E., Yapi- Gnaore C.V., Fantodji T.A. and N'Goran A. 2008. Phenotypical characterization and productive traits of dairy cows from three regions of Cote d'Ivoire. *Zootec Arch* 57: 415-426.
Najimi B., EL Jaafari S., Jlibene M. and Jacquemin J.M. ,2003. Application des marqueurs moteculaires dans l'amëlioration du ыё tendre pour la résistance aux maladies et aux insectes. *B.A.S.E.*, 7, 17- 35.
Ndiaye N.P., Sow A., Dayo G.K., Ndiaye S., Sawadogo G.J., Sembene M. 2015. Genetic diversity and phylogenetic relationships in local cattle breeds of Senegal based on autosomal microsatellite markers. *Veterinary World*, 8: 994-1005. DOI: 10.14202/vetworld.2015.994-1005
Ndumu B.D., Baumung R., Hanotte O., Wurzinger M., Okeyo M.A., Jianlin H., Kibogo H., Solkner J. 2008. Genetic and morphological characterisation of the Ankole Longhorn cattle in the African Great Lakes region. *Genet. Sel. Evol.* 40: 467-490 DOI: 10.1186/1297-9686-405-467
Negrini R., Nijman I. J., Milanesi E., Moazami-Goudarzi K., Williams J. L., Erhardt G.,Dunner S., Rodellar C., Valentini A., Bradley D. G., Olsaker I., Ajmone-Marsan P., Lenstra J. A. , Kantanen J., and The European Cattle Genetic Diversity Consortium 2007. Differentiation of European cattle by AFLP fingerprinting. *Animal Genetics*, 38, 60-66.
Ngono E.P.J., Manjeli Y., Meutchieyië F., Keambou C., Wanjala B., Desta A.F., Ommeh S., Skilton R., Djikeng A. 2014. Genetic diversity of four Cameroonian indigenous cattle using microsatellite markers. *Journal of Livestock Science*, 5: 9-17.
Nijman I.J., Otsen M., Verkaar E.L., De Ruijter C. and Hanekamp E. 2003. Hybridization of banteng (*Bos javanicus*) and zebu (*Bos indicus*) revealed by mitochondrial DNA, satellite DNA, AFLP and microsatellites. *Heredity*, 90: 10-16.
OECD, CEDAO, 2008. Livestock and regional nmirclK' in the Sahel and West Africa: Potentials and dëfis. Capital : Abuja OECD. 163pp.
Oosterhout C.V., William F.H., Wills D.P.M., Shipley P. 2004. Program Note: Microchecker: Softaware for identifying and correcting genotyping errors in microsatellite data. *Mol. Ecol. Notes*, 4: 535-538.
Ouedraogo A., 1989 Contribution a l'etude de la synchronisation des chaleurs chez la femelle Baoule (Bos taurus) au Burkina Faso. These de Doctorat veterinaire. Universite Cheikh Anta Diop, Dakar, 117 pp.
Ouedraogo D., Soudre A., Ouedraogo-Kone S., Zoma B.L., Yougbare B., Khayatzadeh N., Burger P.A., Meszaros G., Traore A., Okeyo M.A.,Wurzinger M., Solkner J. 2020. Breeding objectives and practices in three local cattle breed production systems in Burkina Faso with implication for the design of breeding programs. *Livestock Science,* vol. 232, 103910. Doi :https://doi.org/10.1016/j.livsci.2019.103910
Oumarou A. 2004. "Production laitiere et croissance du zëbu Azawak en milieu reel: suivi et ëvaluation technique a miparcours du projet d^ appui a l^levage des bovins de races Azawak en zone agropastorale au Niger Thëse de Doctorat en Mëdecine Veterinaire, Ecole Inter-Etats des Sciences et Mëdecine Vëtërinaires de Dakar, Universite Cheikh Anta Diop de Dakar, Sënëgal, p. 82.
Paetkau D., Calvert W., Sterling I., Strobeck C. 1995. Microsatellite analysis of population structure in Canadian polar bears. *Mol. Ecol.* 4, 347-354.
Payne W.J.A. 1970. Cattle Production in the Tropics. Volume 1. Longman Group Limited

London. 336 p.
Perez-Pardal L., Royo L.J., Beja-Pereira A., Curik I., Traore A., Fernandez I., Solkner J., Alonso J., Alvarez I., Bozzi R., Chen S., Ponce de Leon F.A., Goyache F. 2010. Y-specific microsatellites reveal an African subfamily in taurine (Bos taurus) cattle. *Anim. Genet*, 41:232241. 10.1111/j.1365-2052.2009.01988.x
Perez-Pardal L., Sanchez-Gracia A., Alvarez I., Traore A., Ferraz J.B.S., Fernandez I., Costa V., Chen S., Tapio M., Cantet R.J.C., Patel A., Meadow R.H., Marshall F.B., Beja-Pereira A., Goyache F. 2018. Legacies of domestication, trade and herder mobility shape extant male zebu cattle diversity in South Asia and Africa. *Sci. Rep.* 8:18027. doi: 10.1038/s41598-018-36444- 7.
Pinart-van der Laan M.H. 2000. The search for QTL using markers: projects and results in sheep and hens. *INRA Prod.* HS, 229-232.
Pinde S., Tapsoba A.S.R., Traore F.G., Ouedraogo R.W., Ba S., Sanou M., Traore A., Tamboura H.H., Simpore J. 2020. Morpho-biometric profiles of the local hen in Burkina Faso. *Int. J. Biol. Chem. Sci.* **14**(6): 2240-2256.
Piry S, Luikart G, Cornuet, JM, 1999. BOTTLENECK a computer programme for detecting recent effective population size reductions from allele data frequencies. *J Hered*, 90. 502-503
Piry S., Alapetite A., Cornuet J.M., Paetkau D., Baudouin L., Estoup A. 2004. GeneClass2: A Software for Genetic Assignment and First-Generation Migrant Detection. *J. Hered*, 95, 536539.
Pitt D., Sevane N., Nicolazzi E.L., MacHugh D.E., Park S.D.E., Colli L., Martinez R., Bruford M.W., Orozco-terWengel P. 2018. Domestication of cattle: two or three events?.events? *EvolAppl*,12(1):123-136. doi: 10.1111/eva.12674.
Planchenault D, 1987. L'elevage In: Elevage et potentialites pastorales saheliennes. Syntheses cartographiques. Burkina Faso = Animal husbandry and sahelian pastoral potentialities. Cartographic synthesis. Burkina Faso. CIRAD-IEMVT - FRA.Wageningen :Wageningen: CTA-CIRAD-IEMVT, 18-22. ISBN 2-85985-121-6
Pritchard J.K., Stephens M., Donnelly P. 2000. Inference of population structure using multilocus genotype data. *Genetics*, 155, 945-959.
Queval R., Petit J.P., 1982. Biochemical polymorphism of FlK'nioglobin in trypanosensitive and trypanotolerant cattle populations and their crosses in West Africa. *Rev. Elev. Med. Vet. Countries Trop.* 35 (2): 137-146.
R Core Team 2013. R: A language and environment for statistical computing. R Foundation for Statistical Computing, Vienna, Austria. URL http://www.R-project.org/.
Randi E., Fusco G., Lorenzini R., Toso S., Tosi G., 1991. Allozyme divergente and phylogenetic relationships among Capra, Ovis and Rupicapra (Artiodactyla, Bovidae). *Heredity* 67, 281-286.
Rannala B. and Mountain J.L. 1997. Detecting immigration by using multilocus genotypes, P. Natl. *Acad. Sci. USA*, 94, 9197-9201.
Raymond M. and Rousset F. 1995. GENEPOP (version 1.2): population genetics software for exact tests and ecumenicism. *J. Hered*, 86, 248-249.
Rege J.E.O., 1992. African animal genetic resources: their characterization, utilization and conservation. In : Rege J.E.O. and Lipner M.E (eds), Proceedings of the Research Plan Workshop held at ILCA, Addis Ababa, Ethiopia, 19-21 February 1992, ILCA (International livestock centre for Africa), Addis Ababa, Ethiopia, 164p.
Rege J.E.O., Aboagye G.S., Tawah C.L. 1994. Shorthorn cattle of West and Central Africa I.

Origin, distribution, classification and population statistics. *World Anim. Rev.* 78:1-14. https://hdl.handle.net/10568/5441

Robertson A. and Hill W.G. 1984. Deviations from Hardy-weinberg proportions: sampling variances and use in estimation of inbreeding coefficients. *Genetics,* 107: 713-718.

Rousset F. 2008. GENEPOP'007: a complete re-implementation of the GENEPOP software for Windows and Linux. *Mol. Ecol. Resour.* 8: 103-106.

RStudio Team, 2020. RStudio: Integrated Development for R. RStudio, PBC, Boston, MA URL *http://www.rstudio.com/.*

Saitou N. and Nei M. 1987. The neighbor-joining method: a new method for reconstructing phylogenetic trees. *Mol. Biol. Evol.* 4: 406-425.

Salifou C F A, Dahouda M, Ahounou G S, Kassa S K, Tougan P U, Farougou S, Mensah G A, Salifou S, Clinquart A and Youssao A K I 2012 Evaluation of carcass traits of Lagunaire, Borgou and Zebu Fulani Bulls raised on natural pasture in Benin. *The Journal of Animal & Plant Sciences*. 22 (4), 857-871.

Sanoga W. A. 2003. Classification, sëllection et dëtermination de la valeur gënëtique des bovins laitiers en ëlevage përiurbain. Mëmoire specialite zootechnie UPB/IDR (Bobo-Dioulasso) . 70 pp.

Sere C. and Steinfeld H. 1996. World livestock production systems: current status, issues and trends. Animal production and health paper No. 127. Rome, FAO. 51pp.

Seydou B., 1981: Contribution a l'etude de la production laitiere du zëbu Azawak au Niger. Faculte de mëdecine et de pharmacie de Dakar) thesis, 102 pp.

Sokouri D.P., Loukou N.E., Yapi-Gnaore C.V., Mondeil F. Gnangbe F. 2007. Caracterisation phënotypique des bovins a viande (Bos taurus et Bos indicus) au centre (Bouake) et au nord (Korhogo) de la Cote d'Ivoire. *Animal Genetic Resources Information*. 40: 43-53

Steffen P, Eggen A, Dietz AB, Womack JE, Stranzinger G, Fries R., 1993. Isolation and mapping of polymorphic microsatellites in cattle. *Anim Genet*, **24**(2):121-4.

Tamura, K., Stecher, G., Peterson, D., Filipski, A., & Kumar, S. 2013. MEGA6: Molecular evolutionary genetics analysis version 6.0. *Molecular biology and evolution*, *30*(12), 27252729. https://doi.org/10.1093/molbev/mst197.

Teriokhin A.T., De Meeus T., Guegan G. F. 2007. On the power of some binomial modifications of the Bonferroni multiple test. *Zh. Obshch. Biol (J. Gener. Biol*), 68 : 332-340.

Thiruvenkadan A.K, Jayakumar V., Kathiravan P., Saravanan R. 2014. Genetic architecture and bottleneck analyses of Salem Black goat breed based on microsatellite markers. *Vet. World,* 7: 733-737.

Traore A., Alvarez I., Fernandez I., Perez-Pardal L., Kabore A., Ouedraogo-Sanou G.M.S., Zare Y., Tamboura H.H., Goyache F. 2012. Ascertaining gene flow patterns in live- stock populations of developing countries: a case study in Burkina Faso goat. *BMC Genetics*, 13:35. doi:10.1186/ 1471-2156-13-35

Traore A., Alvarez I., Tamboura H.H., Fernandez I., Kabore A., Royo L.J., Gutierrez J.P., Ouedraogo-Sanou G., Sawadogo L., Goyache F. 2009. Genetic characterisation of Burkina Faso goats using microsatellite polymorphism. *Livestock Science*, 123, 322- 328. doi:10.1016/j.livsci.2008.11.005

Traore A., Koudande D.O., Fernandez I., Soudre A., Alvarez I., Diarra S., Diarra F., Kabore A., Sanou M., Tamboura H.H., Goyache F. 2016. Multivariate characterization of morphological traits in West African cattle sires. *Arch. Anim. Breed,* 59, 337-344, 2016. doi:10.5194/aab-59-337-2016

Vaiman D., 2000. Etablissement des cartes genetiques. *INRA Prod Anim,* HS, 73-78.
Vaiman D., Barendse W., Kemp S.J., Sugimoto Y., Armitage S.M., Williams J.L., Sun H.S., Eggen A., Agaba M., Aleyasin S.A., Band M., Bishop M.D., Buitkamp J., Byrne K., Collins F., Cooper L., Coppettiers W., Denys B., Drinkwater R.D., Easterday K., Elduque C., Ennis S., Erhardt G., Li L A., 1997. Medium density genetic linkage map of the bovine genome. *Mammalian Genome*. 18: 21-28.
Van Haeringen WA, Den Bieman M, Gillissen GF, Lankhorst AE, Kuiper MT, Van Zutphen LF, Van Lith HA. 2001. Mapping of a QTL for serum HDL cholesterol in the rabbit using AFLP technology. *J Hered* ;92(4):322-6.
Vos P., Hogers R., Bleeker M., Reijans M., Van De Lee T., Hornes M., Frijters A., Pot J., Peleman J., Kuiper M., Zabeau M., 1995. AFLP : a new technique for DNA fringerprinting. *Nucleic Acids Research*, 23, 4407-4414.
Wahlund S.1928. Zusammensetzung von populationenund korrelationsers-chinungen von standpunkt der vererbungslehre ans betrachet. *Hereditas,* 11: 65-108.
Waples R.S. 2015. Testing for Hardy-Weinberg proportions: Have we lost the plot. *J. Hered,* 106, 1-19.
Weber J.L. and Wong C. 1993. Mutation of human short tandem repeats. *Human Molecular Genetics*, 2,1123-1128.
Weir B.S. and Cockerham C.C. 1984. Estimating F-statistics for analysis of population structure. *Evolution,* 38: 1358-1370.
Wendorf F. and Schild R. 1994. Are the early Holocene cattle in the Eastern Sahara domestic or wild? *Evol. Anthropol.* 3: 118-128.
Wintero A.K., Fredholm M., Thomsen P.D. 1992. Variable (dG-dT)n.(dC-dA)n sequences in the porcine genome. *Genomics*, **12**(2): 281-288.
Wolf C., Rentsch J. and Hubner P. 1999. PCR-RFLP Mitochondrial DNA analysis: a reliable mëthod for species identification. *J. Agric. FoodChem*, **47** (4) : 1350-1355.
Wright S. 1951. The genetical structure of populations. *Ann. Eugenics*, 15: 323-354.
Wright S. 1965. The interpretation of population structure by F-statistics with special regard to system of mating. *Evolution,* 19: 395-420.
Wright S. 1969. Evolution and the Genetics of Populations. The theory of Gene Frequencies. University of Chicago Press, Chicago, Vol 2.
Yahaya Z.I., Dayo G.K., Maman M., Issa M., Marichatou H. 2019. Morphobiomëtric cara^risation of the Djelli zëbu of Niger. *Int. J. Biol. Chem. Sci.* **13**(2): 727-744. DOI: http://ajol.info/index.php/ijbcshttp://indexmedicus.afro.who.int.
Youssao I., Tobada P., Koutinhouin B., Dahouda M., Idrissou N., Bonou G., Tougan U., Ahounou S., Yapi-Gnaore C.V., Kayang B. 2010. Phenotypic characterisation and molecular polymorphism of indigenous poultry populations of the species Gallus gallus of Savannah and Forest ecotypes of Benin. *African Journal of Biotechnology* **9**(3): 369-381. DOI : https://doi.org/10.5897/AJB09.1220.
Zerabruk M, Li M-H, Kantanen J, Olsaker I, Ibeagha-Awemu EM, Erhardt G, Vangen O. 2012. Genetic diversity and admixture of indigenous cattle from North Ethiopia: implications of historical introgressions in the gateway region to Africa. *Anim Genet.* 43:257-266. doi: 10.1111/j.1365-2052.2011.02245.x

APPENDICES

Appendix 1: Measurement sheet

PHENOTYPIC MEASUREMENTS

INDIVIDUAL SURVEY FORM PHENOTYPIC MEASUREMENTS

IDENTIFICATION OF THE INVESTIGATOR	
Full name :	**Questionnaire number :**
Date of survey : //	Province :
Village/Site:	**Site code**
Geographical coordinates	
Enclosure : Longitude : Latitude :	**Time of survey**
Park: Longitude: Latitude :	Start : End :

ANIMAL IDENTIFICATION
Identification: Date of birth : ______// Age: Sex : MП FD
Genetic type : Baoule □ Gourounsi □ Zebu □
Date of entry into herd : ______// Method of entry: Birth □ Purchase □
Genetic type of mother: Baoule □ Gourounsi □ Zebu □
Identification of the mother :
Genetic type of father : Baoule □ Gourounsi □ Zebu □
Identification of the father:

HEAD AND HORN MEASUREMENTS

Head length	
Skull length	
Face length	
Head width	
Width of skull	
Face width	
Muzzle circumference	
Length of horns	
Distance Pointe-pointe Corne	
Distance Base-Base Horn	
Ear length	

* Link the питёго of order to the питёго of herd: for example Numero d'ordre 1.3 for 1er herd and 3eme animal in the province.

BODY MEASUREMENTS

Height at withers	
Chest depth	
Height to Sacrum	
Scapulo-Ischial length	
Body length	
Pool length	
Hip width	
Width at ischium	

Tail length	
Thoracic perimeter	
Shoulder room	
Chest width	
Height of hump	
Teat length	
Tape measure weights	
Scale weight	

Printed by Books on Demand GmbH, Norderstedt / Germany